E. W. H. Holdworth

Catalogue of the Birds found in Ceylon

6. Catalogue

E. W. H. Holdworth

Catalogue of the Birds found in Ceylon
6. *Catalogue*

ISBN/EAN: 9783741114342

Manufactured in Europe, USA, Canada, Australia, Japa

Cover: Foto ©Klaus-Uwe Gerhardt /pixelio.de

Manufactured and distributed by brebook publishing software
(www.brebook.com)

E. W. H. Holdworth

Catalogue of the Birds found in Ceylon

6. Catalogue of the Birds found in Ceylon; with some Remarks on their Habits and Local Distribution, and Descriptions of two New Species peculiar to the Island. By E. W. H. HOLDSWORTH, F.L.S., F.Z.S., &c.

[Received February 21, 1872.]

(Plates XVII.–XX.)

In the following Catalogue I have endeavoured to give a correct account of what is known at the present time of the birds resident in and visiting Ceylon. I have included no species about which there appears to be any doubt, except in a few cases; and in those cases I have mentioned the grounds on which their reported occurrence seems likely to be true.

The latest original list of Ceylon birds is that comprised in Mr. E. L. Layard's valuable and generally trustworthy "Notes on the Ornithology of Ceylon" published in the 'Annals and Magazine of Natural History' for 1853–54. Since that time there has been hardly any one in Ceylon who has given systematic attention to the avifauna of the island; and in the preparation of this Catalogue the considerable collection of birds made by myself and Mr. Layard's "Notes" have been the principal materials I have had at my command. I have been able, however, to make use of the extensive knowledge of eastern birds possessed by Lord Walden, the President of this Society, and his large collection of specimens, both of which have been most kindly placed at my service; and they have been of great assistance to me in determining questions of species and nomenclature. I am glad also to acknowledge the help I have received from Mr. J. H. Gurney, Mr. Edmund Harting, Mr. Howard Saunders, and others, in the different groups of birds to which those gentlemen have given their special attention.

Mr. Layard's industry during his eight years' residence in Ceylon resulted in more than doubling the number of birds known in the island ; and he did not leave much for his successors to find out in the low country, where he principally worked. His list included 315 species ; but I have found it necessary to omit a few which appear to have been wrongly identified or to have been recorded by the late Dr. Kelaart on doubtful evidence. I have added 25 species which have been met with during the last few years, and among them two birds hitherto undescribed, making altogether 325 species which apparently have a good claim to be regarded as belonging to the avifauna of Ceylon.

The number of birds not hitherto recognized out of Ceylon is remarkably large considering its small extent of country (equal to only five sixths of the size of Ireland). A few species at one time thought to be peculiar to the island have since been recognized, or are believed to be found, in India or Malacca ; but, so far as is known, the following 37 species are exclusively confined to Ceylon, and are pretty evenly divided between the low country and the hills, most of them, however, being found only in the southern half of the island. They form one ninth of the known Ceylon species.

BIRDS PECULIAR TO CEYLON.

Athene castaneonota.	*Geocichla layardi.*
Tockus gingalensis.	*Merula kinnisi.*
Palæornis calthropæ.	*Oreocincla spiloptera.*
Loriculus indicus.	*Alcippe nigrifrons.*
Chrysocolaptes stricklandi.	*Drymocataphus fuscocapillus.*
Brachypternus ceylonus.	*Pomatorhinus melanurus.*
Megalaima zeylanica.	*Layardia rufescens.*
—— *flavifrons.*	*Garrulax cinereifrons.*
Xantholæma rubricapilla.	*Rubigula melanictera.*
Centropus chlororhynchos.	*Drymoipus validus.*
Phænicophaës pyrrhocephalus.	*Zosterops ceylonensis.*
Prionochilus vincens*.	*Cissa ornata.*
Eumyias sordida.	*Temenuchus senex.*
Erythrosterna hyperythra.	*Eulabes ptilogenys.*
Buchanga minor.	*Munia kelaarti.*
—— *leucopygialis.*	*Palumbus torringtoniæ.*
Dissemurus lophorhinus.	*Gallus stanleyi.*
Brachypteryx palliseri.	*Galloperdix bicalcarata.*
Arrenga blighi.	

* Recently discovered by Mr. Legge, R.A. A notice of this species is given in a postscript to this Catalogue.

Excepting *Phœnicophaës* and *Prionochilus*, which are quite Malay forms, all these peculiar Ceylon species belong to genera found in India. Most of these genera range from India more or less to the countries east of it; and the nearest allies of *Cissa ornata* are almost confined to Eastern Asia.

I have not included *Batrachostomus moniliger* or *Kelaartia penicillata* among the species peculiar to Ceylon, as they are believed to be found also in South India; and I have likewise omitted *Malacocercus striatus*, as I much doubt its distinctness from *M. malabaricus*.

Geographical distribution.—In this Catalogue I have given approximately the geographical range of most of the species found in Ceylon, from which it will be observed that all those not peculiar to the island are, with very few exceptions, known in India; the majority of them extend to Burmah, many of them to some of the Malay islands and China, and a few to Australia. *Goisachius melanolophus* is a remarkable example of a common Malaccan species having four times been found in Ceylon, and, strangely enough, only on the west side of the island, although it has not yet been observed on the adjoining Indian continent. A converse example exists in *Hirundo hyperythra*, of which one or two specimens have been brought from Malacca, that species being otherwise considered quite peculiar to, as it is abundant in, Ceylon.

The Ceylon birds which range to the westward of India belong to species of generally wide distribution, and consist principally of Raptorial, Grallatorial, and Natatorial forms; the exceptions being examples of *Hirundo*, *Cypselus*, *Halcyon*, *Ceryle*, *Cuculus*, *Cisticola*, and *Pyrrhulauda*.

Of the species which extend to Australia those belonging to *Calobates*, *Strepsilas*, and *Terekia* are of very wide distribution; the Ceylon species of *Haliaëtus*, *Excalfactoria*, *Charadrius*, *Ægialitis*, and *Mycteria* have a considerable range east and south-east of India; and *Attagen minor* and *Sterna gracilis* seem alone to be, so far as is known, especially Australian.

Indian families absent from Ceylon.—The *Vulturidæ*, *Eurylaimidæ*, *Pteroclidæ*, *Otididæ*, *Glareolidæ*, *Gruidæ*, and *Mergidæ*, all families included in the Indian avifauna, have no recognized representatives in Ceylon. Of the *Vulturidæ*, one species breeds so far south as the Neilgherries; but Ceylon agrees with the Indian archipelago and the countries south of continental Asia in having no Vulture. The *Eurylaimidæ* have their stronghold on the eastern side of the Bay of Bengal and in the Malay islands; and a representative of this family may yet be found in Ceylon. It is also not improbable that stragglers of the common South-Indian species of *Pteroclidæ* and *Otididæ* may one day be met with in the north of the island. *Glareola* may likewise be looked for; but the *Gruidæ* and *Mergidæ* are not likely to range so far south.

Position and Character of the Island.—Without entering into the question of whether Ceylon was originally a continuous portion of India or formed part of a lost Malay continent, as believed by the late Sir J. Emerson Tennent, it may be desirable to point out the

principal features of the country as it now stands. Its position (between 6° and 10° N. lat.) is almost equatorial. Practically it is an
island about 35 miles (at its least distance on the extreme north)
from India, increasing to nearly 60 miles at the connecting sandbank
of Adam's Bridge, and to about 150 miles between Colombo and
Cape Comorin. It possesses the character of a true oceanic island
in having deep water (no bottom at 150 fathoms) within a very few
miles of the land all round the coast, excepting only between Adam's
Bridge and Point Pedro, the parts of the island nearest to India. The
water shoals abruptly on the south side of Adam's Bridge, and has
only a depth of a few fathoms north of it until it passes the line
between the north point of Ceylon and the nearest part of India,
whence it gradually deepens into the Bay of Bengal. Adam's Bridge,
the narrow connecting-link between Ceylon and India, and said to be
of comparative recent formation, consists of sandstone covered with
loose sand, which is alternately beaten up on and removed from the
north and south sides by the sea and wind during the successive north-
east and south-west monsoons. It terminates on the Indian side in
the island of Ramisseram, between which and the continent is the well-
known Paumben Channel. On the Ceylon side the bridge ends in the
island of Mannar, which is separated from the mainland by a considerable expanse of shallow water or mud banks, according to the state of
the tide, with a narrow winding channel deep enough for the passage
of small native vessels. The bridge itself has also several narrow
openings or "scours" at different parts, so that, although Ceylon is
virtually connected with India by means of Adam's Bridge, it may be
regarded as practically distinct, and, as might be expected, it has
species peculiar to itself in all the great divisions of the animal king-
dom. Its length is 271 miles and its greatest breadth 137 miles.

For ornithological purposes Ceylon may be divided into two parts
—the northern and southern halves, the northern portion being, with
the exception of a few isolated hills, entirely low country; this is
continued throughout the maritime districts of the south; and the
whole coast is surrounded by a narrow belt of sandy beach. The low
country generally is extensively laid out with paddy-fields; but there
are large tracts in the northern half of the island which are still in
the normal condition of forest, or, from the poverty of the soil or
the scarcity of rain, are only occasionally cultivated, and support a
scattered growth of bushy jungle rarely attaining the character of
forest. This last was the nature of the country round Aripo, where
I spent a good deal of time and obtained a great number of the
commoner birds. At the north and on the north-east side there are
large lagoons or backwaters, the resort of countless Waders; and there
and on the inland lakes or tanks (as they are generally called) Ducks
and Terns of various kinds are abundant in winter, and many other
birds at all seasons. The avifauna of the northern half of the island
is quite Indian in its character. The east and south-east parts also
contain a good deal of wild country; they are thinly populated, and
are visited the least by Europeans. One district is the home of the
few remaining Veddahs, the supposed aborigines of Ceylon, who,

although almost savages compared with the rest of the natives, are said to retain the honourable distinction of high caste. The extreme south and south-west are generally well cultivated; and paddy-fields and cocoa-nut plantations are general in that part of the country. The mountain-districts lie almost in the centre of the southern half of the island; and in this half, at various elevations ranging from sea-level to 8000 feet, are found by far the greater number of the peculiar Ceylon birds. A conspicuous feature of the Ceylon hills is the luxuriant vegetation which clothes them from their foot to the tops of the highest ranges; and although masses of rock may be seen here and there projecting from the mountain-sides, even these are largely covered with ferns and creeping plants. The mountain region may be divided ornithologically into the lower and upper hills. The country up to between 1500 and 1600 feet, the elevation of Kandy, is only partly cultivated; and its diversified character provides suitable habitats for a great variety of birds. This is particularly the case in the neighbourhood of Kandy, where there is some really wild jungle, in which some of the rarer hill species as well as low-country birds are found at certain seasons.

From the elevation of Kandy to about 5000 feet are the coffee-districts; and where this cultivation is general the number of birds is small, and they are found mostly at the higher and lower boundaries of the estates. If, however, the soil be unsuited for coffee and the jungle remain uncleared, birds are numerous, and many of the peculiar kinds, *Athene castaneonota*, *Palæornis calthropæ*, &c., may be met with. These lower hills are the great resort for the passerine immigrants; and birds of prey abound there. From 5000 to 8200 feet (the highest point in the island) constitutes what I shall have frequent occasion to speak of as the upper hills. They are almost entirely covered with tree jungle, with a dense undergrowth of "nilloo" (*Strobilanthes*), small straggling bamboo, tree ferns, and a variety of other plants. These hills are the great stronghold of the Sambur Deer; and Elephants and Leopards mount to their summits. Nuwara Eliya, the sanatorium of the island and a place where I have collected largely, is at an elevation of 6000 feet, and lies in a narrow plain, the houses being mostly scattered along the sides at the foot of the surrounding jungle-covered hills. The birds found in this locality and the neighbouring district are not numerous in species; but they are mostly of kinds peculiar to the island, and include *Chrysocolaptes stricklandi*, *Brachypteryx palliseri*, *Cissa ornata*, *Zosterops ceylonensis*, and several others, whose range does not generally extend far below the upper hills.

Migratory Birds.—The migration of birds within and into Ceylon is a subject about which there is still a great deal to be learnt; but, owing to the absence of observers, there is little reason to expect much trustworthy information will be gained for some time. The migrations take place at the changes of the monsoons. The S.W. monsoon blows steadily and for the most part strongly from April to October on the west side of the island. In October there is a lull for a few days between the two winds. It is the season for cyclones in the

Bay of Bengal; and then very often there is stormy weather on the
Ceylon coast.   At the first decided indication of the N.E. monsoon
setting in, the true migratory birds begin to appear; they are generally first seen in the north and north-west of the island, and gradually extend over the western side and to the hills.   At the same
time there is a large influx of resident species to these parts of the
country, which during this N.E. monsoon are less exposed to the
violence of the wind than in the other season.   There is no positive
evidence whence these birds come; but I think there can be little
doubt that it is from the eastern side and some parts of the central
districts.   Many of them are certainly found there during the S.W.
monsoon; but no continuous observations have been made on the
eastern side, and there is little known of what resident Ceylon species
are to be met with there at any particular season.   It being a great
game country, Europeans who visit the eastern jungles devote their
time more to sporting than to natural history.   I may give an instance
showing that there must be a good deal yet to be done in certain parts
of the country.   In February 1871 I obtained at Nuwara Eliya two
specimens of a Flycatcher (*Erythrosterna hyperythra*) of which the
type specimen in the Berlin Museum, obtained in 1866, was the only
one known; it came from the Ceylon hills; but that species is certainly
not found in the hill districts during a great part of the year, and
yet it has not been observed elsewhere*.   Towards the close of
February the N.E. monsoon comes to an end, and is followed by five
or six weeks of fine calm weather before the usually stormy burst of
the S.W. monsoon.   The migratory birds now take their departure,
and many species resident in the island leave its western side.   A
Tern, however, in immature plumage and believed to be *Sterna gracilis* has only been observed on the Ceylon coast in summer; but as
the Ceylon summer is at the same time as the Australian winter, the
fact of this Tern being found at Colombo in July is an additional
reason for believing it to be that Australian species.   I have also
only seen Frigate-birds during the summer; but Mr. Layard has
recorded their occurrence in February.

With respect to the breeding-season for Ceylon birds it is difficult
to fix any definite rule.   The climate in the low country is always
hot and damp, and birds of some species or other are nesting
throughout the year.   In many cases the breeding-time appears to
depend on the monsoons; but I believe it often varies with the same
species in different parts of the island.   On the upper hills, where
there is the nearest approach to a cold season of any part of Ceylon,
and where the midday tropical heat is succeeded by cold nights and,
in January and February, by severe frost, the breeding-season follows
the rule in temperate climates and usually begins about April; in
other parts of the country either nesting or moulting appears to be
always going on.

From what I have said of the character of the country it will be
evident that Ceylon possesses, in its swamps, jungles, forests, rivers,
and coasts, the conditions suitable for the existence of a great variety

* See No. 127, footnote.

of birds; the waters on the coast and inland swarm with fishes; the country is alive with insects and reptiles; and vegetation is most luxuriant in its growth; food of all kinds abounds; and there is no winter in the low country. It is no wonder, therefore, that species and individuals are numerous; but although I have, I believe, been able to add something to the good work done by Mr. Layard, the subject is yet far from being exhausted, and much remains to be done in examining the eastern side of the island generally, in discriminating many of the wading birds, and in working out the Terns and other birds found on the coast.

In this Catalogue I have followed Jerdon's arrangement of the species as given in his ' Birds of India,' and have adopted the names he uses, except in a few cases where older titles may be more properly employed.

1. FALCO PEREGRINUS, Gmelin.

Europe, Asia.

2. FALCO PEREGRINATOR, Sund.

Ceylon, India, W. Asia.

3. HYPOTRIORCHIS CHICQUERA, Daud.

Ceylon, India.

These species are recorded by Layard as found in Ceylon; but they have not come under my notice.

3 bis. HYPOTRIORCHIS SEVERUS, Horsf.

In a collection of birds sent home by Mr. S. Bligh, and consisting entirely of hill species shot by himself in one of the coffee-districts, is an undoubted specimen of the Indian Hobby, an unexpected addition to the list of Ceylon birds. It was killed whilst hawking after dragonflies.

Ceylon, N. India, Malacca, Java, Philippines.

4. TINNUNCULUS ALAUDARIUS, Gmel.

The Kestrel is widely distributed in Ceylon; I have seen it, however, most frequently in the northern part of the island, and a pair of these birds for many weeks frequented a small clump of cocoa-nut palms near my house at Aripo. I have also observed it at Nuwara Eliya during winter; and it is often met with in the coffee-districts. Although probably a migrant, it certainly spends several months in Ceylon.

Europe, Asia.

5. ASTUR TRIVIRGATUS, Temm.

This is a hill species, and not very uncommon. I have examined specimens of the bird in Ceylon, and have now before me a very good one killed by Mr. Forbes Laurie.

Mountains in Ceylon and India.

6. MICRONISUS BADIUS, Gmelin.

Said by Layard to be very common and widely distributed in Ceylon. I have identified several specimens of it.
Ceylon, India, Burmah, Malaya, Hainan.

7. ACCIPITER VIRGATUS, Temm.

I have a specimen of this Sparrow-Hawk from the lower hills. Layard does not mention this species ; but it may possibly have been the one recorded by Kelaart as *A. nisus*, which I have no reason to think has been found in Ceylon.
Bill dark bluish ; irides yellow ; feet yellow.
Ceylon, India, Burmah, Malaya, Formosa.

8. AQUILA PENNATA, Gmelin.

Recorded by Layard.
Ceylon, India, W. Asia, N. Africa, S. Europe.

9. NEOPUS MALAIENSIS, Reinw.

Tolerably numerous in the hill country, and well known in the coffee-districts. I have seen several skins in different states of plumage, which were obtained from the hills around Kandy.
Ceylon, India, Burmah, Malaya.

10. NISAËTUS BONELLI, Temm.

Recorded by Layard as having been obtained by the late Dr. Templeton, R.A.
Ceylon, India.

11. LIMNAËTUS CRISTATELLUS, Temm.

This noble bird, mentioned by Layard under the name of *Spizaëtus limnaëtus*, Horsf., is well known in the hill country, and not unfrequently visits the poultry-yards of the coffee-planters. I have seen it on many occasions at Nuwara Eliya, and listened to its squealing cry as it soared in wide circles over the plain. In the beginning of 1871 I procured a fine living specimen, and shipped it at Colombo for the Society's Gardens ; but it died soon after the vessel sailed. The feet and claws in this species are very powerful.
Mountainous parts of Ceylon and India.

12. LIMNAËTUS NIPALENSIS, Hodgson.

Recorded by Layard as having been procured by the late Dr. Kelaart on the hills, at an elevation of 4000 feet. It is remarkable that this species should be found in Ceylon, as in India it is only known from the northern hills; but Mr. Blyth tells me that he identified Dr. Kelaart's specimen, and has no doubt of its being the true *L. nipalensis*, Hodgs. No other example of this bird has been recognized in Ceylon.
Ceylon, N. Indian hills, Formosa, Japan.

13. SPILORNIS BACHA, Daudin.

Generally distributed over the island, frequenting trees on the margin of tanks and marshy places in the low country, and near open grass-land among forest-jungle on the hills. One specimen, which I shot near Aripo, disgorged a Tree-snake (*Passerita*) more than 3 feet long and nearly uninjured. Another, obtained at Nuwara Eliya, fell to the shot as if mortally wounded, although only slightly injured in one wing; it soon recovered, and became sufficiently tame to feed from my hand. I was fortunately enabled to bring the bird with me to England; and it is now alive in the Society's Gardens.

S. spilogaster, Blyth, from Ceylon, is now recognized as the immature condition of *S. bacha*; and there is no doubt that the *Hæmatornis cheela*, recorded by Layard as common in Ceylon, may also be referred to the same species.

Bill dusky ; irides golden yellow ; cere, legs, and feet dull yellow.
Ceylon, S. India, Andamans, Malaya.

14. PANDION HALIAËTUS, Linn.

Rare in Ceylon, and I have only seen it on one occasion ; it was perched on a buoy in Galle Harbour ; and I was able to watch it from a short distance for a considerable time. Lord Walden has two specimens of it from Ceylon.
Europe, Asia, Africa.

15. POLIOAËTUS ICHTHYAËTUS, Horsf.

This Eagle I have only seen in the north of the island, where it is not uncommon near the coast. I shot an immature specimen at Aripo in November 1866. The irides were brown, but Jerdon states (App. B. of Ind. iii. p. 869) that in the adult bird they are pale yellow.
Bill black ; "irides pale yellow ;" feet yellowish white.
Ceylon, India, Burmah, Malaya.

16. HALIAËTUS (CUNCUMA) LEUCOGASTER, Gmelin.

This is the common Sea-Eagle of Ceylon, and is probably found all round the island, although I do not remember having observed it at the extreme south. It may occasionally be seen soaring over Colombo Harbour and the adjoining lake ; but further north, in the neighbourhood of Aripo and Mannar, several pairs of these noble birds may generally be found, each generally in its own district, and rarely wandering far away. In the strait separating the island of Mannar, at the east end of Adam's Bridge, from the mainland of Ceylon, the narrow channel for the passage of boats in the midst of the expanse of shallow water around is marked here and there with stakes ; and on these may generally be seen perched one or two pairs of this Eagle, and sometimes a pair of *Polioaëtus ichthyaëtus*. As the receding tide lays bare the extensive banks of soft mud on each side the Eagles keep a sharp look-out for the crabs, which are abundant just at the edge of the water, and, pouncing on their prey, sail

off to some favourite tree, where the hard shell of the crab is broken
up and the animal devoured.   One of these stations, further down
the coast, was on the cross-trees of a government flagstaff at Aripo;
and the ground below was always littered with crab-shells and fish-
bones, the remains of many a meal provided from the refuse of the
fishermen's nets, which were hauled in on the beach close by.   Sea-
snakes (*Hydrophis*) are said to be a favourite food of this species; and
these reptiles are abundant on the Pearl-Oyster banks nine or ten
miles off the Aripo coast; but I have never observed the Eagles so
far from the land.

A curious instance came to my knowledge of this bird having
apparently thriven on most unnatural food.   My friend, Dr. Boake,
the late Principal of Queen's College, Colombo, once pointed out to
me an example of this Eagle of full size, but in immature plumage.
It had been recently brought to him by a native, who said he had
reared the bird in his own hut.   In answer to an inquiry as to what
he had fed the bird on, he said "rice and curry."   This is the
universal food of the natives; and dogs and cats appear to thrive as
well upon it; but that a Sea-Eagle should have been reared on such
food seemed incredible.   However, the matter was soon tested by a
supply of rice and curry being given to the bird; and the statement
of the native was quickly confirmed by the rapid disappearance of
the whole of the food.   The next day some fish was given, and the
Eagle, once having tasted it, could never afterwards be induced to
touch rice and curry.

In a male example of this Sea-Eagle which I shot at Aripo I
found the liver of an enormous size, covering the whole of the pectoral and a great part of the abdominal regions.

Bill dusky blue; cere yellow; irides brown; feet yellowish white.

Ceylon, India, Burmah, Malaya, Australia.

16 *bis*. BUTEO DESERTORUM, Daudin.

Lord Walden has received a single specimen of this Buzzard from
Ceylon.

Ceylon, India, Persia, S. Europe, Africa.

17. CIRCUS SWAINSONII, A. Smith.

Common in the Aripo district throughout the year; and I have
frequently seen it at Nuwara Eliya in July and August.   The pale
rump of the brown birds attracts attention as they hunt backwards
and forwards over the open country.

Bill black; irides yellow; feet yellow.

Asia, Africa.

18. CIRCUS CINERACEUS, Montagu.

I have only identified this species on one occasion; it was killed
near Colombo.   Although probably not uncommon in Ceylon, it is
certainly not so numerous there as the last species.

Bill black; irides yellow; feet yellow.

Europe, Asia, Africa.

19. CIRCUS MELANOLEUCOS, Forster.

This species was first described from, and has since been identified as a visitor to Ceylon; but I have never met with it. Layard procured it on the west coast.

Ceylon, India, Tientsin.

20. CIRCUS ÆRUGINOSUS, Linn.

This Harrier is probably only an occasional visitor to Ceylon. I observed a pair of these birds near Aripo in January 1870; and after several ineffectual attempts to get near them, I at last succeeded in shooting the female, a handsome specimen, with grey wings and tail. Layard does not appear to have met with this species; but it is included doubtfully in the list of birds in Tennent's 'Natural History of Ceylon.'

Bill black; irides and cere yellow; feet deep yellow.

Europe, Asia.

21. HALIASTUR INDUS, Bodd.

Common on the coast, especially on the northern half of the island. Specimens in various states of plumage were obtained at Aripo. I have also seen it at Colombo and Trincomalie.

Ceylon, India, Borneo.

22. MILVUS GOVINDA, Sykes.

This bird has very much the same habits and distribution in Ceylon as the last species. Neither of them, however, frequents the towns so much as they both do in India. In early morning at Aripo I have seen a flock of fifty or sixty Pariah Kites, in company with about a dozen of the other species, eagerly clutching at and feeding on the winged Termites which were rising in a cloud from an ant-hill not far from my house. The Crows were busily engaged on the same work, but kept at a respectful distance, apparently not liking to join in the general scramble going on among their more powerful neighbours, the Kites.

Ceylon, India, Burmah, Malaya, Andamans, China, Formosa, Hainan.

Note.—There are two, perhaps three, closely allied species of Kite found in India, the smallest of which, Mr. Gurney tells me, is identical with M. affinis of Australia; and there is some doubt as to which is best entitled to the specific name of govinda. As it is not quite clear to which of these the Ceylon birds belong, the above geographical range may not be strictly correct.

23. PERNIS PTILORHYNCHUS, Temm.

Given by Jerdon as P. cristata, Cuvier. I had an opportunity of seeing this bird alive in Ceylon; and Mr. Forbes Laurie has recently shown me a good specimen which he shot on the hills. Lord Walden has also received examples of it from Ceylon. Mr. Laurie's specimen agrees pretty closely in dimensions with those given by

Jerdon ; but Mr. Gurney tells me that the birds from Ceylon are usually larger than those from India. Although this bird is well known from Ceylon, it appears not to have been hitherto recorded from that island.

Ceylon, India, Burmah to Malaya.

24. BAZA LOPHOTES, Cuv.

Not very numerous, but has been found both in the low country and on the hills. I have seen specimens from the Kandy district.

Ceylon, India.

25. ELANUS MELANOPTERUS, Daud.

I have only seen specimens of this handsome bird from the hills, where locally it is not uncommon. Layard obtained it in the low country.

Ceylon, India, part of Africa, S. Europe.

26. STRIX INDICA, Blyth.

Formerly included in *S. javanica*, De Wurmb., which Jerdon has recently (Ibis, 1871) stated to be more nearly allied to *S. candida*, Tickell. *S. indica* is very local in Ceylon, and is entirely confined to the north of the island. Layard gave the fort of Jaffna as the only locality for it ; but I have since obtained it at Aripo, where a pair of these Owls were resident. They frequented a government storehouse in my compound, each regularly perching in a dark corner under the roof, at opposite ends of the long building, and apparently living in harmony with the hundreds of Bats which hung from the roof and walls around. I have never observed these birds out of doors until some time after sunset.

Bill horny yellow ; irides black ; feet yellowish brown.

Ceylon, India.

27. SYRNIUM INDRANEE, Sykes.

This bird is found in the low country in the northern half of the island and on the lower hills ; but although well known to and dreaded by the natives as a bird of ill omen, it does not appear to be anywhere numerous. Doubts have been expressed as to whether the so-called "Devil-bird" is really an Owl ; but I have frequently questioned the native hunters about the bird, which is so notorious in Ceylon for its horrible cries ; and they have described it in such terms as to leave no doubt in my mind about its being an Owl, and probably of this species.

I have only seen specimens of it from the Kandy district ; but it has been found in several parts of the island, and I once had an opportunity of hearing the bird under very favourable circumstances near Aripo. I was lying out in wild jungle about eight miles from my house, and five from the nearest native village, watching for Bears. It was bright moonlight ; the Nightjars had long ceased their churring notes, and there was an almost unnatural stillness around—the midnight silence of the jungle, only occasionally broken by the distant roar of the surf. For several hours I had been watching the small drinking-hole in front of me; and it was now time for the Bears to come if they meant to visit the pool at all that night. I was eagerly scrutinizing the openings among the bushes, when piercing cries and convulsive screams suddenly issued from a small patch of bushy jungle about thirty yards on the left of my hiding-place. My hunter at first thought a Leopard was there, and told me to keep quiet; but the cries increased, and became so horribly agonizing, that it was difficult to believe murder was not being committed; so, jumping up with my double rifle in my hand, I ran cautiously down to the patch of jungle, my trusty servant following with a second gun. Before I reached the place all was as silent as before, and the idea of the Devil-bird flashed across my mind. This was afterwards confirmed by the hunter, who, however, did not apparently care to talk much about it. A careful examination of the sandy ground among and around the bushes when daylight appeared resulted in no evidence of any tracks of Leopards or recent traces of other quadrupeds. I have no doubt, therefore, that it was this dreaded Owl which had disturbed our night watch; and although my sport was spoilt for the night, I did not regret having heard for once the really appalling cries of this ill-omened bird. The dimensions of a Ceylon specimen are:—Length 20 inches, wing 13, tarsus 2.

Ceylon, S. India, Malacca, Formosa (*Swinhoe*).

28. HUHUA PECTORALIS, Jerdon.

Some three or four years ago, whilst I was in Ceylon, Mr. Samuel Bligh brought to me for identification some specimens of a Horned Owl, which appeared to us, after examination, to be identical with *Huhua nipalensis*, Hodgson, except in being smaller, but agreeing in that respect with the measurements of a bird from S. India described by Jerdon as *H. pectoralis*. Considerable confusion has existed between *H. nipalensis*, Hodgs., from Nepal, *H. pectoralis*, Jerdon, from S. India, and *H. orientalis*, Horsfield, from Java; and the subject is referred to by Jerdon (B. of Ind. vol. i. p. 132) as a matter on which "materials are wanting to form a just conclusion." Jerdon has since (Ibis, 1871, p. 346) stated his opinion that the Nepal species will stand, and has united the other two under *H. orientalis*—but, I understand, in the absence of a specimen from S. India for direct comparison.

A comparison of one of the Ceylon birds with specimens of true *H. orientalis* and *H. nipalensis* in the British Museum has satisfactorily shown, however, that they are three very distinct species, and that the Ceylon bird is very probably the same as *H. pectoralis* from S. India. In this conclusion I am supported by Mr. Gurney and Lord Walden.

H. pectoralis may be described as like *H. nipalensis*, but very much smaller, both of them wanting the *closely* barred plumage of *H. orientalis*. It is, I think, evident from Jerdon's measurements of *H. nipalensis* that they were taken from a Malabar specimen of what was supposed to be that species; but he says in his description, the brown bars of the under parts "in some tending to coalesce and form a pectoral band." In his figure of *H. pectoralis* in the 'Madras Journal' great prominence is given to this band; but in the Ceylon bird it is not very distinct; and, as Mr. Gurney has pointed out to me, this difference appears to be owing to the light intervals between the dark transverse bars on the pectoral feathers being not so light in the plate as in the Ceylon bird*. The following are the comparative measurements of the three species :—

	Length. in.	Wing. in.	Tarsus. in.
H. nipalensis, from Nepal (B.M.) ..	28–29	18·5	2·5
H. pectoralis, from Ceylon	22	16	2
H. orientalis, from Java (B.M.)	20	12	1·5

H. pectoralis is not uncommon on the lower Ceylon hills, and has probably been mistaken, without much critical examination, for the common *Ketupa ceylonensis*.

Bill yellow; irides brown; feet dull yellow.

Ceylon, S. India.

29. KETUPA CEYLONENSIS, Gmel.

Generally distributed over Ceylon, but perhaps more common in the low country than on the hills. I have frequently met with them near Aripo. Large trees overhanging a tank are a favourite resort of these birds, and I have often found them in the early morning perched day after day on the same branch. They are frequently captured and kept alive by the natives.

Bill dusky yellow; irides yellow; feet dirty yellow.

Ceylon, India, Burmah, China; Palestine (*Tristram*).

30. EPHIALTES BAKKAMUNA, Forster.

Some difficulty exists in determining how many species of small Tufted Owl are found in Ceylon, partly on account of the confusion there has been among the species or races found in India, and variously named by different naturalists, and partly because there is some doubt about the correctness of Dr. Kelaart's identification of the species he records. There is, I think, no question, however, that the very common and widely distributed species is that given by Jerdon as *Ephialtes lempigi*, Horsf., but described from Ceylon in 1781 by Forster as *Strix bakkamuna*, an unfortunate name, as it is evidently meant for "bakha muna"—lit. "Fish-Owl," and the Singhalese name for *Ketupa ceylonensis*. Forster's plate, however, shows that his bird was the common *Ephialtes*.

E. bakkamuna is very common in most parts of the low country, and is also found about Kandy and on the lower hills. It was a

---

* Since the above was written Mr. Bligh has sent home a specimen of this Owl for the Norwich Museum. It is generally rather darker, and probably more mature than the one in my possession; and the pectoral band is very distinct, leaving no doubt of the validity of Jerdon's species.

constant evening visitor to the trees surrounding my house at Aripo; and its single call of *whock*, repeated at short and regular intervals, was frequently to be heard far into the night. It is a bird of rapid flight. A young bird of this species was completely tamed by Mr. Bligh in Ceylon; it would fly to his finger, and delighted in being stroked and played with; and this tameness continued undiminished after the bird had become adult. I have often had this amusing little pet in my hands.

The dimensions of a Ceylon specimen, a female, are :—Length 8 in., wing 6·5, tarsus 1·4, tail 3.

Bill dusky; irides yellow; feet greyish.

31. EPHIALTES SUNIA, Hodgson.

Kelaart mentions, under this name, "a very small reddish-yellow Eared Owl, occasionally seen in the very highest parts of the island." I have some recollection of seeing a specimen from the hills, which I believe was the bird he referred to, and I think the species may be included in the Ceylon list. It is probably the rufous phase of *E. pennatus*, Hodgs.

I have no evidence of any other species of *Ephialtes* in Ceylon than the two I have here given.

32. ATHENE CASTANEONOTA, Blyth.

Peculiar to Ceylon. It is probably confined to the southern half of the island, and has been killed on the hills and in the low country, but is by no means common. I obtained two specimens that were killed in the neighbourhood of Kandy. It is admitted as distinct from *A. castanoptera* from Malaya.

Bill yellow; irides ——?; feet greenish brown.
Ceylon.

33. NINOX HIRSUTA, Temm.

Rare in Ceylon; I have only seen one specimen, which was obtained in the central district. Layard also only met with it once in the course of eight years.

Bill green; irides yellow; feet dingy yellow.
Ceylon, India, Borneo, China, Japan.

34. HIRUNDO RUSTICA, Linn.

Referred to by Layard under *H. gutturalis*, Scop., the Indian representative of the European species; but the grounds for separating them appear to be of the slightest description, and I shall adopt the now general opinion that they are the same. This bird is a winter visitor to Ceylon, and generally distributed, but especially abundant in the low country. Most of these birds are young ones, without the long tail-feathers.

Asia, Africa, Europe.

35. HIRUNDO DOMICOLA, Jerdon.

Confined to the upper hills in Ceylon. It is a very familiar bird,

commonly nesting in the verandas at Nuwara Eliya and in the district.
Ceylon, Neilgherries, Malaya.

36. ? HIRUNDO DAURICA, Linn.

Layard records having obtained a single specimen at Point Pedro. It is more probable, however, that this bird was *H. erythropygia*, Sykes, the S. Indian species, which had not at that time been distinguished from *H. daurica*, a northern bird and having a wide range to the eastward. They both have the under plumage streaked.
Ceylon, S. India.

37. HIRUNDO HYPERYTHRA, Layard.

This Swallow was discovered by Layard in 1849, and until recently was considered peculiar to Ceylon; but I have seen a specimen lately received by Lord Walden from Malacca, and it has been otherwise recorded from that country. It is abundant in the central and, at times, in the western and southern districts of the island, both in the low country and on the lower hills; but I have not observed it at Nuwara Eliya or in the north. Its distinguishing character consists in the whole of the underparts being deep chestnut.
Bill black; irides brown; feet black.
Ceylon, Malacca.

38. ACANTHYLIS GIGANTEA, Temm.

This bird is said to be well known at Nuwara Eliya; and Layard mentions hearing of the native report that it breeds in hollow rhododendron trees; but there is probably some mistake, as I could hardly have failed to notice the bird under such circumstances. I have only seen it from the coffee-districts; and although specimens have undoubtedly been obtained at Nuwara Eliya, I expect it will be found to be only an occasional visitor there.
Ceylon, S. India, parts of Malaya.

39. CYPSELUS MELBA, Linn.

Probably a winter visitor to Ceylon. It is found in some abundance on the hills at that season, but is rather local in its distribution. I have seen it at Nuwara Eliya in the cold season; and it remains there several months, particularly frequenting some precipitous cliffs overlooking the plain on which the little town is built. In the afternoon fifty or sixty of these birds might any day be seen on the wing dashing past the hill-sides in pursuit of insects, or sweeping in wider circles at a considerable elevation.
Hill-regions in Ceylon, India, W. Asia, Africa, Europe.

40. CYPSELUS AFFINIS, Gray.

Layard speaks of this bird as migratory, and breeding in April in large numbers about the rocks at Damboul. I have also found it nesting, but in August, under the rocks overhanging the entrance to the famous temple at Damboul; and as it breeds in Ceylon during

the summer months, I have no doubt it is a resident species. It
has been met with in other parts of the island, but is local. I have
not observed it on the upper hills.

Ceylon, India.

41. CYPSELUS BATASSIENSIS, Gray.

Very common in the low country and particularly abundant in the
north, where the palmyra is the common palm, on which it builds
its nest. I have not observed this bird at Nuwara Eliya, but have
known the following species sometimes mistaken for it there.

Bill black; irides brown; feet brownish.

Ceylon, India, Assam, Burmah.

42. COLLOCALIA FUCIPHAGA, Thunberg (1772).

There are several localities in Ceylon in which this little Swift has
been known to breed; and Layard has given a good description of one
which he visited. These breeding-stations are at various elevations,
from close to the sea to the highest hills. Kelaart states that he has
"heard from very good authority that some years ago baskets of the
edible nests were obtained from a cave on the Pedrotallagalla hill.
Very nutritious soup was made of them for the invalids who at that
time resided at Nuwara Eliya." This hill is the highest (8200) in
Ceylon and overlooks the still all-important sanatorium; but I could
not ascertain the situation of the cave. The birds, however, are
numerous at Nuwara Eliya in winter; and I have obtained specimens
there, as well as in the low country between Colombo and Kandy.

Bill black; irides dark brown; feet purple-brown.

Ceylon, N. and S. India, Assam, Malay islands.

43. DENDROCHELIDON CORONATUS, Tickell.

Layard says this species is generally distributed; but I have only
seen it between Colombo and Kandy and in the south. I shot a pair
of these birds near Colombo at the end of May, and have seen it in
abundance in Kandy itself in March. I think there is little doubt
that it is a resident, although perhaps migrating from one part of the
island to another. I have neither seen nor heard of it at Aripo or
Nuwara Eliya.

Bill black; irides dark brown; feet bluish black.

Ceylon, S. and Central India, Pegu.

44. BATRACHOSTOMUS MONILIGER, Blyth.

I have only seen one skin of this curious bird. It was procured
by Mr. H. Nevill close to Amblangodde Lake, a few miles north of
Galle. I was shown the spot where it was killed—a small piece of
recently cleared land nearly surrounded by rather low jungle. The
species has been but rarely met with, and, so far as is known, is confined to the south-west of the island, in the country lying between
Adam's Peak and Galle.

Mr. Blyth tells me that Jerdon's description of this species was

taken from a Ceylon specimen, but that the one from S. India is
probably the same.
Ceylon, S. India?

45. CAPRIMULGUS KELAARTI, Blyth.

This species, first discovered in Ceylon, is entirely confined to the
hills, ranging from about 2000 feet upwards, and common in the
Nuwara Eliya district.   It is very noisy during March and April, at
the commencement of the breeding-season, appearing with great
regularity a few minutes after sunset from its accustomed hiding-place
in the thick jungle.   I have reason to think this Nightjar leaves the
upper hills during the cold season and descends to a more temperate
climate.
Bill dusky; irides dark brown; feet fleshy brown.
Ceylon, Neilgherries.

46. CAPRIMULGUS ATRIPENNIS, Jerdon.

Tolerably common near Colombo and in the south; I have also
obtained it in the interior, about twenty miles from Trincomalie.   It
is, I believe, a low-country bird; and I have not met with it above
the foot of the hills.
The specific distinctness of this Nightjar was hardly ascertained
when Layard published his "Notes on the Ornithology of Ceylon;"
and when he speaks of so rare and remarkably coloured a species as
C. mahrattensis, Sykes, being "abundant in the vicinity of Colombo
and throughout the southern province," and that bird has not been
met with in Ceylon by any subsequent collector, it is not unreason-
able to conclude that the species intended is that which I have since
identified as C. atripennis from the same localities.
Bill dusky; irides brown; feet pale brown.
Ceylon, S. India.

47. CAPRIMULGUS ASIATICUS, Latham.

Common in the low country, especially in the northern half of the
island, where it is resident.   I have found it breeding in September
at Aripo, its two eggs being deposited on a bare sandy spot under the
shelter of a bush.   At Colombo it is numerous in the cinnamon-gardens during at least part of the year, hiding during the day at the
foot of the bushes; but I have no recollection of seeing this bird in
the south of the island.   Ceylon specimens are very grey compared
with those from India, a good series of the latter which I have
examined having all a conspicuous rufous tinge on the upper surface.   This is only observable in Ceylon birds in young specimens.
There is another point in connexion with this species to which I wish
to direct attention; and it may be desirable to extend it to other species
of Caprimulgidæ.   Jerdon and other authors have been accustomed
to group the species of Caprimulgi in accordance with the number of
tail-feathers which have a white terminal spot, this spot being supposed to be found only in the male.   I need hardly say that it is only
too common for collectors to omit any notice of the sex of the birds

they shoot; and this frequent omission has no doubt led to the over-looking of the fact that in *C. asiaticus* the female has the white tail-spot as well as the male, although about one fourth shorter. This is not a peculiarity of birds from Ceylon, as in a series of Indian specimens of unknown sex in Lord Walden's collection I was able to separate them at once into two groups agreeing on this point pre-cisely with the known sexes of the birds of my own collecting. I first noticed the presence of the spot in the female in a specimen I shot at Aripo in 1866, and I made a note of it; but the skin was de-stroyed by rats; I have since obtained two more examples showing precisely the same character, so that the first could not have been an accidental variation. I have not been able to ascertain whether the same character is found in the two other Ceylon species, as all my specimens of them are males.

Bill dusky, tip darker; irides very dark; feet brownish flesh.
Ceylon, India, Burmah.

48. HARPACTES FASCIATUS, Forster.

Only found among wild tree-jungle in the southern half of the island. I have seen it about twelve miles from Colombo, in a wild uncultivated district in the low country, and also at Nuwara Eliya, in February; but it is not very commonly met with, and is perhaps somewhat local in its distribution. In its manners it resembles the Flycatchers, and has generally a peculiar fluttering mode of flight.

Bill dark blue; cere smalt; irides brown; feet lavender.
Ceylon, South and Central India.

49. MEROPS VIRIDIS, Linn.

Exceedingly abundant in the northern part of Ceylon, where it is a resident. It is also found sometimes at Colombo and on other parts of the coast. Whilst living at Aripo I had constant opportunities of observing these birds closely, as the railings of my veranda were a favourite perching-place for them, and they would allow me to ap-proach within a few feet without showing any alarm. Forty or fifty of these beautiful birds generally roosted in a small bushy tree only a few yards from the house. This species seems to prefer a low sta-tion when looking out for its prey, frequently perching on a small stick only a few inches from the ground. The Ceylon birds gene-rally have the blue throat which is found in the variety described as *M. torquatus* by Hodgson.

Bill black; irides blood-red; feet lead-colour.
Ceylon, India to China.

50. MEROPS PHILIPPINUS, Linn.

A migratory bird; generally distributed over the low country, but, like the preceding species, very numerous in the north. I have first observed it at Aripo at the end of September; and it remains there till the change of the monsoon in April. It is a noisy bird, with a lofty, dashing flight, successfully pursuing the dragonflies, and then sailing back on outstretched wings to its favourite station on the dead branch of some neighbouring tree, where the insect is killed and swallowed.   In the early mornings of March, when there has been but little wind stirring, and the sea was as smooth as glass, I have frequently observed these Bee-eaters hunting for insects close to the surface, and a quarter of a mile from the shore.   I have noticed this bird frequently at Colombo, but only in small parties.   At Aripo I have often seen sixty or seventy on the same tree; in fact, during its stay in Ceylon, it is more numerous there than the resident species.

Bill black; irides blood-red; feet lead-colour.

Ceylon, India, Burmah, Malaya.

51. MEROPS QUINTICOLOR, Vieill.

This is a hill-species, and a resident in Ceylon.   I have shot it in August at the foot of the hills in the south, and I have frequently seen it on the lower hills in the neighbourhood of Kandy; but it is nowhere so numerous as either of the preceding species, and is generally seen singly or in pairs.   I have not observed it on the upper hills.   Of two Ceylon specimens, with the chestnut border to the black throat-band, one has the tail entirely green, and the other with the central feathers blue.

Bill black; irides blood-red; feet lead-colour.

Ceylon, India, Burmah.

52. CORACIAS INDICA, Linn.

This bird, although undoubtedly locally common in the north of Ceylon, has never come under my notice in the jungly district of Aripo; nor have I seen it in the south of the island.   In the country between Colombo and Kandy, however, I have frequently met with it; and its often noticed habit of perching on the top of a bare pole or the stem of a dead tree is also characteristic of the bird in Ceylon.   Its flight is regular and crow-like; but when perched its head is sunk on its shoulders, giving the bird a remarkably clumsy appearance, as is also the case with the Bee-eaters when not on the wing.

Bill blackish; irides dark brown; feet fleshy yellow.

Ceylon and the greater part of India.

53. EURYSTOMUS ORIENTALIS, Linn.

Layard met with three examples of this bird; but it has never come under my notice.

Ceylon, northern half of India, Burmah, Malaya, China.

54. PELARGOPSIS GURIAL, Pearson.

The synonymy of this bird has been much confused.   It is mentioned by Layard under the name of *Halcyon capensis*, Linn., and is described by Jerdon under the heading of *H. leucocephala*, Gmel.   Mr. R. B. Sharpe (P. Z. S. 1870), however, has worked out the question of identity, and has restored to it Pearson's original name of *gurial* (1841), by which it should be known.

I have seen skins of this species, but have not met with the bird alive. Layard speaks of its frequent occurrence on the east side of Ceylon, and also of its being found about Caltura, on the west coast. The latter locality is, I have heard, a good one for this bird ; and I have reason to think it is also sometimes met with on the lower hills.

Bill red ; irides brown ; feet dull red.

Ceylon, India eastward.

55. HALCYON SMYRNENSIS, Linn.

H. fusca, Bodd., is now admitted as identical with the old Linnean *H. smyrnensis*. In Ceylon this Kingfisher is abundant in the low country wherever there is water, frequenting alike the neighbourhood of paddy-fields and the banks of rivers. It is perhaps less numerous in the north than elsewhere, but it was not uncommon at Aripo. Ceylonese specimens are generally more brightly coloured than those from other countries.

Bill deep red ; irides brown ; feet vermilion.

Ceylon, India to China, Asia Minor.

56. HALCYON PILEATA, Bodd.

Recorded by Layard, under *H. atricapilla*, Gmel., as having been killed by him in the north of the island. It appears to be an eastern species, and rare both in India and Ceylon.

Ceylon, India, Burmah, Malaya, China.

57. CEYX TRIDACTYLUS, Pall.

Widely distributed in Ceylon, but nowhere common, and only to be procured with difficulty. I have never seen the bird alive, but at various times obtained three specimens, which were killed in the central district.

Bill coral-red ; irides brown ; feet red.

Ceylon, India, Malaya.

58. ALCEDO BENGALENSIS, Gmel.

Common everywhere in Ceylon. It is always to be found at Nuwara Eliya, as well as in all parts of the low country.

Bill reddish, with the upper part dusky ; irides brown ; feet coral-red.

Ceylon, India to China, Malaya.

59. CERYLE RUDIS, Linn.

This is a common species, frequenting rivers more than tanks or paddy-fields. I have frequently met with it at Aripo ; and it is particularly abundant on the southern rivers.

Bill black ; irides brown ; feet brown.

Ceylon, India to S. China, Malaya, W. Asia, S. Europe, Africa.

60. HYDROCISSA CORONATA, Bodd.

Confined to wild forest jungle in the central and northern parts of the island. I have seen it occasionally a few miles from Aripo; and whilst travelling through an extensive tract of forest on the road between Kandy and Trincomalie, small parties of these birds were frequently observed on the tops of the trees, or slowly sailing across the road from one part of the forest to the other. In the early mornings their harsh cries mingled discordantly with the howlings of Monkeys (*Presbytes*), the call of the Jungle-fowl, and the more musical notes of the Long-tailed Robin (*Kittacincla*), almost the only sounds to be heard in this primitive jungle, far from the borders of cultivation, and only disturbed by occasional travellers or the bell of the light-stepping postal runner.

Ceylon, S. India.

61. TOCKUS GINGALENSIS, Shaw.

Considerable confusion has existed between this species and *T. griseus*, Latham; and it is desirable to mention that the species properly known as *T. gingalensis* is only found in Ceylon. Under the above heading Jerdon has inadvertently spoken of both in his 'Birds of India,' but he has since corrected the mistake (Ibis, 1872, p. 5).

Tockus gingalensis is, according to Layard, not uncommon in certain districts; and Lord Walden has received several specimens of it. It keeps, I believe, mostly to the forests; and I have only once obtained it at Aripo, where its harsh cry betrayed its presence on a low tree close to my house. The colour and shape of the bill in this bird vary a good deal with age.

Bill yellowish, more or less marked with black; irides reddish brown; feet slate grey.

Ceylon.

Tockus griseus, Lath., is said by Jerdon to be also found in Ceylon; but I cannot hear of any well authenticated specimens.

62. PALÆORNIS ALEXANDRI, Linn.

Bill red; irides buff; feet slate.

Ceylon, India, N. Burmah.

63. PALÆORNIS TORQUATUS, Bodd.

Bill red; irides buff; feet slate.

Ceylon, India, W. Asia, Tropical Africa?

These two species are exceedingly abundant in the north of Ceylon; but I have not seen them on the hills or in the south.

64. PALÆORNIS ROSA, Bodd.

I have only met with this species in the southern parts of Ceylon, where it is very destructive to the grain crops; but it is also found at times on the lower hills generally. I have seen a flock of fifty of these birds fly down one after another to a field of paddy; and each biting off a ear of the green corn, return to a neighbouring tree to

devour the plunder; and this has been repeated again and again.
The above three species are constantly caged by the natives; and
few native dwellings are without one or other of these favourite pets.
Bill yellow above, black below; irides buff; feet greyish.
Ceylon, India.

Note.—There is some doubt about the further range of this
species, a closely allied form, with yellow under wing-coverts, having
probably been confounded with it.

65. PALÆORNIS CALTHROPÆ, Layard.

Peculiar to Ceylon. It was first obtained by Layard at Kandy,
where it is frequently numerous; and it is said to be generally distributed over the hills. Although recorded by Kelaart from Nuwara Eliya, I suspect this beautiful bird is only a rare visitor to that
cool region, as I have never seen a Parrot of any kind at that elevation, and I have always been on the look-out for this species in
particular. The colouring in the sexes is alike, except that in the
female the green on the side of the head is less distinct, and the bill
is black instead of red.
Bill, ♂ red, ♀ black; irides buff; feet slate.
Ceylon.

Note.—As some confusion has existed with regard to the correct
spelling of the specific name of this species, I may mention, on the
direct authority of Mr. Layard, that it was given from " Calthrop,"
a family name.

66. LORICULUS INDICUS, Gmel.

Peculiar to Ceylon. The history of this species has been fully
discussed in a paper by Lord Walden (Ibis, 1867, p. 467), from
which it appears that, although Edwards first figured and described
the bird, it should stand as *L. indicus*, Gmel., according to the rules
of zoological nomenclature. The name is unfortunate, as it is certain that the species is not found out of Ceylon; but it was not
known by the earlier writers from what part of the Dutch settle-
ments the bird described by Edwards was obtained.
This little bird is common in many parts of the southern half of
Ceylon, and particularly quite in the south. It frequents cultivated
ground and large native gardens; and I have sometimes seen it on
the cocoanut-trees busily biting off and apparently eating the chip-
like flowers. I believe it is confined to the low country. It is often
caged by the natives, and, like allied species, sleeps suspended from
the top of its cage by its strong curved claws. There is little difference in the colouring of the sexes; but individuals vary a good deal
in the extent and brilliancy of the golden gloss on the back.
Bill reddish orange above, orange below; irides white; feet dull
yellow.
Ceylon.

67. PICUS MAHRATTENSIS, Latham.

Common in the Aripo district, and, so far as I know, only found

in the north of Ceylon. It appears to frequent low jungle, and I
have rarely seen it except on dead wood near the ground and old
fences. It is a resident species.

Bill slate; irides dull red; feet lead-colour.

Ceylon, India.

Picus macei, Vieill., has been recorded by Dr. Kelaart as being
found in Ceylon; but I think its occurrence is very doubtful, in
which opinion I am strongly confirmed by Mr. Blyth.

68. YUNGIPICUS GYMNOPHTHALMOS, Blyth.

This little Woodpecker was discovered in Ceylon by Layard, and
it is said to have been since found in S. India. It frequents the
upper branches of large trees, and, although generally running over
the bark in true Woodpecker fashion, may sometimes be observed
perched across the smaller twigs. I have only obtained it in the
neighbourhood of Colombo; but it is also found in the south.

Bill greenish slate; irides pale buff; orbital skin purple; feet
greenish slate.

Ceylon, S. India?

69. CHRYSOCOLAPTES FESTIVUS, Bodd.

By the kindness of Lord Walden, I am enabled to include this
handsome Woodpecker in my list of Ceylon birds. The two specimens, male and female, in his collection, are labelled "November
1865, Cocarry." The name is probably that of a small native
village in the north-west of the island, not far from the Aripo
district, as I have reason to know that the birds collected at that
date for Lord Walden were procured not many miles from where I
was afterwards residing. Future collectors in Ceylon, who are not
familiar with this species (described by Jerdon under the name of *C.
goensis*, Gmel.), may recognize it by its black back and golden wings,
the underparts being coloured much as in *C. stricklandi*, Layard.

Ceylon, parts of South and Central India.

70. CHRYSOCOLAPTES STRICKLANDI, Layard.

Peculiar to Ceylon, and confined to the hills. It is abundant at
Nuwara Eliya and in all tree-jungle in that district, ranging from
the forest-clad Pedrotalagalla (8200 feet), the highest point in the
island and overlooking the Nuwara Eliya plain, through the coffee-
districts, to the Kandy country. The female has the whole top of
the head and crest black, spotted with white; and a young bird of
that sex had the lower part of the back black, faintly barred with
white, with crimson feathers appearing among the others: the bill
in this bird was only two thirds the length of that in the adult.

Layard states that the irides of this species are red-brown; but I
think he must have been mistaken, as in four specimens I obtained
at Nuwara Eliya, and which I myself prepared, the irides were buff,
those of the young bird being rather paler than the others.

Bill greenish white; irides buff; feet greenish slate.

Ceylon hills.

71. CHRYSOPHLEGMA CHLOROPHANES, Vieill.

I have only procured this species at the foot of the hills in the south; but it has been also obtained in other places much nearer Colombo. When not feeding, it is in the habit of stationing itself on the highest branch of a dead tree, and there repeating its peculiar note, which has little of the harsh sound so generally characteristic of the Woodpeckers.

Bill slate, with the base yellow; irides dull red; feet dull green.
Ceylon, S. India.

72. MICROPTERNUS GULARIS, Jerdon.

Two specimens of this Woodpecker were procured by me a few miles from Colombo. Although decidedly a scarce species, and I shot these two birds in January and July, they were both killed in native gardens not a quarter of a mile apart. Layard met with it in the south; and I have seen one or two skins from the central district.

Ceylon specimens have the lower parts rather darker than those from India. Layard gives this bird under the name of *M. phaioceps*, Blyth.

Bill lead-grey; irides red-brown; feet slaty brown.
Ceylon, S. India.

73. BRACHYPTERNUS AURANTIUS, Linn.

Recorded by Layard as very abundant in the Jaffna peninsula in the north of the island. I occasionally saw at Aripo what may have been this species, and heard its remarkable cry, but failed to procure a specimen.

73 *bis.* BRACHYPTERNUS PUNCTICOLLIS, Malh.

A specimen of this bird has been quite recently received by me from the western side of the island. A further examination of the Golden-backed Woodpeckers found in Ceylon therefore appears desirable, as the species generally met there is more likely to be *B. puncticollis*, common in Southern India, than *B. aurantius*, which has a more northerly range. *B. puncticollis* may be recognized by its white-dotted throat and under neck.

Ceylon, S. India.

74. BRACHYPTERNUS CEYLONUS, Forster.

Peculiar to Ceylon; not uncommon near Colombo, but very numerous in the south. Dr. Kelaart* says it is "found in great abun-

---

* The results of my own collecting at Nuwara Eliya and in the neighbouring jungles during almost every month in the year oblige me frequently to receive with suspicion the notices by the late Dr. Kelaart of the occurrence of birds, and of their abundance, in that district. The subjects to which Dr. Kelaart gave his special attention were mammals and reptiles, and in these he did good work; but ornithology was a very subordinate study with him, and he rarely, if ever, used a gun.

.dance at Nuwara Eliya;" but I have never seen it on the hills, and I have no doubt that *Chrysocolaptes stricklandi*, another red-backed Woodpecker already noticed, was mistaken by Dr. Kelaart for this species. This bird especially frequents the cocoanut-trees, and is a conspicuous object as it works its way by rapid jerks up the slender trunks of these palms. The natives in the south call it the "Toddy-bird," and say it visits the palms for the sake of the toddy, which is largely collected in that and some other parts of the island. The insects feeding on the toddy are no doubt the real attraction. Its principal food is ants, as is the case with all the low-country Wood-peckers, their stomachs being always found more or less crammed with these ubiquitous and troublesome insects. Both sexes have the red occipital crest; but the male has the top of the head sprinkled with the same colour, whilst the female has that part spotted with white.

Bill slate; irides red; feet pale greenish.
Ceylon.

75. Megalaima zeylanica, Gmel.

Peculiar to Ceylon. This bird is closely allied to *M. caniceps*, Franklin, and is noticed under that name by Layard; but it is a smaller bird, with the anterior portion of its plumage much browner, and the lighter markings reduced in size and distinctness. It is common in the low country, except in the north. I have never seen or heard it in the Aripo district; and it does not ascend above the lower hills. The flight of this bird is straight, but rather heavy. It feeds on berries, and may be often seen clinging to the smaller twigs on the outside of a tree whilst eating the fruit which grows at their extremities.

Bill dull orange; irides brown; orbits yellow; feet yellow.
Ceylon.

76. Megalaima flavifrons, Cuvier.

Peculiar to Ceylon. It is not confined to the hills, as stated by Layard, but is exceedingly abundant even close to Colombo, and ranges from near the coast to an elevation of 5000 feet. It is the only Barbet I have seen so high; and I have not observed it there except during the N.E. monsoon, a time at which there is a great influx of migratory birds and of low-country species to the hills. I have not seen it in the north; and it is not so numerous as the last species in the extreme south of the island. At the village of Ile-neratgodde, about 17 miles from Colombo, in a district abounding with native gardens, cocoanut-topes, and paddy-fields, and where I have collected a great variety of birds, the air used to resound with the loud notes of this and the preceding species of Barbet, a partial silence only occurring for an hour or two during the extreme heat of the day. *M. flavifrons* is a more sprightly bird than *M. zeylanica*, and can be readily distinguished from it when on the wing.

Bill horny yellow; irides red-brown; feet dark grey.
Ceylon.

77. XANTHOLÆMA INDICA, Lath.

Layard speaks of this bird under the name of *Megalaima philippensis*. It is confined to the north. I have only met with it at Aripo, where it is found throughout the year. Perched on some dead branch near the top of a tree, with its throat swelling and its head bowing at the utterance of each note, this handsome little Barbet repeats its monotonous cry of *poohp, poohp, poohp* for half an hour at a time, with only occasional intervals of a minute or so. Whilst thus engaged it changes the direction of its head with every note; and to this I think is mainly due the often noticed variation in the sound; but the range of direction is a full semicircle; and after often listening to the bird from different positions, I have no doubt that the voice is also dropped a little when the head is turned quite on one side. In Ceylon, as in India, this bird is known by the name of " coppersmith ;" and that title is also applied about Colombo to the following species.

Bill black ; irides red-brown ; feet pink.
Ceylon, India, Burmah, Malaya.

78. XANTHOLÆMA RUBRICAPILLA, Gmel.

Peculiar to Ceylon, and common in the low country in the southern half of the island. I have also frequently seen it at Trincomalie ; and Layard has procured it at Jaffna ; but I have never met with it in the Aripo district. It is very common about Colombo. The note of this bird is very much like that of *X. indica*, but is not nearly so loud, and is repeated quickly four or five times without a pause ; then resting for three or four seconds, the bird goes on as before. The call of *X. indica* sounds like distinct heavy blows of a hammer on a copper vessel heard in the distance ; that of *X. rubricapilla* like a series of light taps on the same metal.

Bill greyish black ; irides red-brown ; feet pink.
Ceylon.

In a young bird I obtained in July near Colombo the bright colours about the head and neck were not developed, except a small patch of orange below the eye and a tinge of yellow on the forehead. The bill was dark grey ; irides pale brown, and feet dusky flesh.

79. CUCULUS CANORUS, Linn.

Recorded by Layard as found in Ceylon. He obtained one example of it near Colombo; but I have not met with it.
Asia, Africa, Europe.

80. CUCULUS SONNERATII, Latham.

Kelaart procured several specimens ; and I have seen it from near Colombo and the lower hills.
Ceylon, S. India.

81. CUCULUS MICROPTERUS, Gould.

This Cuckoo was recorded by Dr. Kelaart as a mountain species ;

but the only two examples I met with were obtained in half-cultivated land in the low country near Colombo. To this species may probably also be referred a bird closely resembling *C. canorus* which I watched for some time in an English garden at Colombo a few days after my arrival in Ceylon.

Bill yellowish, dusky above ; irides pale yellow ; feet yellow.
Ceylon, India, Burmah, Malaya, China.

Layard has described a Ceylon Cuckoo under the name of *C. bartletti*; but there is some doubt about what the bird is. Jerdon places it under *C. poliocephalus*, Lath., which, however, has not been recognized in the island ; it may be *C. sonneratii*.

82. HIEROCOCCYX VARIUS, Vahl.

This bird is probably a migrant from India. Layard procured it near Colombo ; but I have only met with it on the hills at Nuwara Eliya, in the beginning of the year.

Bill greenish yellow, dusky above ; irides yellow ; feet yellow.
Ceylon, India, Burmah, Malaya.

83. POLYPHASIA PASSERINA, Vahl.

Referred to by Layard as *Cuculus tenuirostris*, Gray, and by Jerdon (B. of India, vol. i. p. 333) as *P. nigra*, apud Blyth. Jerdon has lately, however (Ibis, 1872, p. 14), gone more into the nomenclature of the species, and placed it under the above heading. It is migratory to Ceylon, but appears much later than most of the other visitants. Layard gives February for its arrival about Jaffna ; but I have first seen them at Aripo in the beginning of January, and then they all at once became abundant, frequenting low bushes in the jungle, and ranging in colour from dark grey to completely rufous on the upper parts. No two specimens were exactly alike ; but all were of some shade of grey beneath, and more or less barred. The rufous-bellied species is an eastern bird, and unknown in Ceylon.

Bill black above, red-brown below ; irides hazel ; feet dull yellow.
Ceylon, India.

84. SURNICULUS DICRUROIDES, Hodgson.

Resident, but rather a scarce bird in Ceylon. It has been found on the lower hills, near Kandy ; and I have obtained specimens in immature and adult plumage in the low country near Colombo, and in the extreme south of the island. Although at first sight this Cuckoo may be readily mistaken for a King Crow, having the same general colour and remarkable shape of tail, it is not difficult to distinguish it when within a moderate distance. It usually perches lower and alights more frequently on the ground, besides having little of the Flycatcher-action so common among the *Dicruri*.

Bill black ; inside of mouth deep orange ; irides dark brown ; feet black.
Ceylon, India, Burmah.

85. LAMPROCOCCYX MACULATUS, Gmel.

This beautiful Emerald Cuckoo was first made known from Ceylon, and appears to be the one given by Kelaart and Layard in their Catalogue (1853) under the name of *Cuculus xanthorhynchos*, Horsf., a Malay species. I have seen no specimens of it; and it is undoubtedly rare.

Ceylon, India.

86. COCCYSTES JACOBINUS, Bodd.

C. melanoleucos, Gmel.; Jerdon, B. of Ind. no. 212.

Common in the north of the island. These birds are always numerous in the Aripo district, frequenting bushes and low trees, and usually perching on the highest branches. In December and January (the commencement of the breeding-season with many birds in Ceylon) they are very noisy and incessantly flying from one place to another, one or more males apparently chasing the female, and uttering their clamorous cries. Layard mentions finding a young Cuckoo of this species under the care of a pair of Mud-birds (*Malacocercus*); and, from the frequent battles I observed between this Cuckoo and a pair of *Malacocercus striatus* which were nesting in a low tree close to my house, I have no doubt that the Black-and-white Crested Cuckoo frequently lays its eggs in the nest of that common Babbler.

Bill black; irides red-brown; feet lead-colour.

Ceylon, India, Africa.

87. COCCYSTES COROMANDUS, Linn.

I believe this handsome Cuckoo is very scarce in Ceylon. I have only seen two specimens, both from the Kandy district.

Bill black; irides reddish brown; feet lead-colour.

Ceylon, India, Burmah, Malaya.

88. EUDYNAMIS HONORATA, Linn.

Formerly known as *E. orientalis*, Linn.

Layard says of this bird in Ceylon:—" Wherever Crows are found, there the Coël is found also." I have only seen this bird, however, during the N.E. monsoon, from November to April. During this period it is very common in the Aripo district; and I have also found it numerous near Colombo. After April, I have never met with the species until towards the end of the year. I believe it is a true migratory bird. Among the specimens I have shot in January and February is a young male in the spotted plumage, but having the top of the head rusty brown; in other respects the colours are the same as, but purer than, those in the female. These Cuckoos are very noisy in the morning and evening.

Bill dull green; irides crimson; feet slate-colour.

Ceylon, India, Burmah, N. Malaya, S. China.

89. ZANCLOSTOMUS VIRIDIROSTRIS, Jerdon.

This is a low-country species, and, so far as I know, not extending

to the south of the island. It is found abundantly throughout the
year in the north; and I have occasionally met with it near Colombo.
It is skulking in its habits, creeping rapidly through the low bushes,
and rarely exposing itself when it has once been alarmed.

Bill apple-green; irides deep red, orbits cobalt; feet dark leaden.
Ceylon, S. India.

90. Phœnicophæus pyrrhocephalus, Forst.

This Cuckoo has hitherto been found only in Ceylon. It inhabits
tree-jungle in the low country near the foot of the hills. One specimen, alive but injured, was brought to me by some natives who had
caught it only a few miles from Colombo. I saw a second flying
across a road in the Central Province, and followed it for some distance through the jungle, but failed to obtain it. Its flight was
weak; but it moved rapidly through the trees, half flying and half
hopping from branch to branch. Layard says the irides of this
Cuckoo are white; but in the living bird (a male) I had they were
brown, and they are marked as of that colour in specimens in Lord
Walden's collection.

Bill light apple-green above, bluish green below; irides brown;
orbital skin crimson; feet dark leaden.
Ceylon.

91. Taccocua leschenaultii, Less.

I am indebted to Mr. Forbes Laurie for the opportunity of examining a male specimen of this fine Cuckoo, hitherto unknown in
Ceylon. He tells me it was obtained in the Doombera valley (1800
feet), not far from Kandy, that it came into his hands immediately
after it was shot, and he himself prepared the skin. As a S.-Indian
species it is likely to occur in Ceylon; and the Doombera valley is
a wild district, from which I have known many of the rare and peculiar Ceylon birds to have been obtained.

Bill red, tip yellow; irides reddish; feet lead-colour.
Ceylon, S. India.

92. Centropus rufipennis, Illiger.

Common generally in the low country. It was very abundant at
Aripo, feeding very much on the ground, where there was always a
large supply of grasshoppers.

Bill black; irides red; feet black.
Ceylon, India, Burmah, Malaya.

93. Centropus chlororhynchus, Blyth.

Peculiar to Ceylon, and, I believe, almost confined to the lower
hills in the Central district. I have only seen this bird alive on
one occasion, and then in thick jungle under trees. It is either
very scarce or escapes notice from its skulking habits. From C.
rufipennis it may always be distinguished by its green bill, if

not by the very rich purple gloss over the anterior portion of its plumage.
Bill pale green; irides red; feet black.
Ceylon.

94. NECTAROPHILA ZEYLONICA, Linn.

Common in the low country. I have frequently seen it in the gardens at Colombo; but have not met with it at Aripo. Layard speaks of it as abundant in the southern and midland districts.
Ceylon, India.

95. NECTAROPHILA MINIMA, Sykes.

I do not remember seeing this bird in the Aripo district, although Layard states that it is common in the north of the island. It is occasionally seen at Colombo.
Ceylon, S. India.

96. ARACHNECHTHRA ASIATICA, Latham.

This species was very common at Aripo, and was found there at all seasons. I have also seen it in the south. At a Government rest-house in the extreme south of the island, where I was staying in August 1869, a pair of these birds had a nest in the veranda; it was fastened to the end of an iron rod hanging from the roof and once used for suspending a lamp. The birds showed very little fear, although I was for several days sitting within a few feet of the nest, engaged in the preparation of specimens. I have obtained this species at Nuwara Eliya in October.
Bill black; irides red-brown; feet black.
Ceylon, India, N. Burmah.

97. ARACHNECHTHRA LOTENIA, Linn.

This is a very common species at Colombo, and is said by Layard to be plentiful in the southern and midland districts. I have no note of its occurrence at Aripo. Some specimens have the bill very much curved.
Bill black; irides brown; feet black.
Ceylon, S. India.

98. DICÆUM MINIMUM, Tickell.

I have procured this little bird in all parts of the island; and specimens obtained at Nuwara Eliya were precisely the same as those from Aripo and elsewhere.
Bill flesh-colour; irides brown; feet fleshy brown.
Ceylon, India, Burmah.

99. PIPRISOMA AGILE, Tickell.

Layard records having obtained a pair of these birds on the Central road.
Ceylon, India.

100. DENDROPHILA FRONTALIS, Horsf.

Layard speaks of this bird as "abundant about jack-trees," which are only found in the low country. Although I have known it killed in such parts of the island, I have always considered it a hill species, as it is one of the common birds at all seasons at Nuwara Eliya and on the upper hills. Jerdon states that in India it is most abundant on the Neilgherries—a situation corresponding in a remarkable manner with the higher hills in Ceylon, the birds and plants of the two ranges being in most respects the same.

These little Nuthatches appear to keep in small parties at all times of the year, and are very active in examining the branches of any trees they may happen to visit. The colours of this bird soon lose their brightness after death; and the peculiar delicacy of the tints can hardly be discovered in a cabinet specimen.

Bill coral-red ; irides golden ; feet yellow-brown.
Ceylon, India, Assam, Burmah, Malaya.

101. UPUPA NIGRIPENNIS, Gould.

Very abundant in the Aripo district during the winter months, and occasionally in the summer. Some of these birds are no doubt residents in Ceylon ; but their numbers in the north are largely increased about October, either by migrants from India or from the east side of the island. Layard speaks of it under the name of *U. senegalensis*, Sw., and says he "shot young birds, not fully fledged, in August." This would agree with the breeding-time of the Hoopoe in Burmah, of which Jerdon says :—"I found it breeding in holes of trees in June and July." If Layard's birds, however, were bred in Ceylon, as might be supposed from his statement that they were not fully fledged, then there are two distinct breeding-seasons for this species in the north of the island, as in January 1870 I found, in my compound at Aripo, a nest of the Hoopoe in a hole in a small mustard-tree (*Salvadora persica*). I caught the old bird as it was leaving the nest ; and after enlarging the hole, came down to three young birds, just hatched, and resting on a bed of rotten wood. These nestlings were quite naked, and their bills were barely a quarter of an inch long.

The Hoopoe was found by Layard on the east and south-east coasts, and once at Colombo. I have also had a specimen from the neighbourhood of Kandy.

The flight of this bird is easy and undulating ; and its note is repeated whilst it is on the wing, as well as when perched on the top of a tall bush.

There is some variation in the colours and dimensions of the Hoopoes found in Ceylon, the tendency being towards the characters of the Burmese variety described by Jerdon. Of three specimens shot at Aripo, one has the bill at front 1·9 inch, the closed wing 5 ; first primary entirely black, chin whitish, and the feathers of the posterior half of the crest white between the black and rufous. The last characters have been regarded as specially belonging to

U. epops; but in this specimen there is no white spot on the first primary. This bird was killed in December. In a second specimen, shot in February, the bill is 2·25 inches, closed wing 5·1, the general colour of the bird very rufous, and a white spot on the left first primary, but only a very minute speck on the right. A third, a male killed in December, agrees generally with the Indian form, has no spot on the first primary, but has the bill at front 2·15 inches, and the closed wing 5·25. The last and largest of these specimens did not exceed 10·8 inches in length; and they may all doubtless be referred to *U. nigripennis*, if that form be really distinct from *U. epops*.

Bill black, base flesh-colour; irides brown; feet dark leaden.

Ceylon, India, Burmah?

102. LANIUS ERYTHRONOTUS, Vigors.

Very common in the Aripo district and in other parts of Northern Ceylon. I have also seen it occasionally in the Cinnamon Gardens at Colombo; but it does not appear to visit the hills. A cup-shaped nest of this species was built in a thorn-bush close to my house at Aripo; but the young birds had left it before my arrival there in the beginning of April. In a subsequent year I obtained young birds able to fly as early as the middle of February, and older ones nearly full-grown in March. These young birds were all very rufous, with the head, upper back, and flanks closely barred, the lower part of the back more broadly marked, and the secondaries rufous with their centres dusky. Layard says "the young are fledged in June;" but they are out some months earlier than that in the Aripo district. These birds feed very much on dragonflies and grasshoppers.

Bill black; irides dark brown; feet black.

Ceylon, India, Central Asia.

103. LANIUS CRISTATUS, Linn.

Referred to by Layard as *L. superciliosus,* Linn. This bird I have found common in the north, west, and central parts of Ceylon during the winter months. It remains about Aripo from October to April, and is tolerably common at Nuwara Eliya during the same period. Layard mentions their being particularly numerous at Hambantotte, on the south-east coast, but does not say at what time of the year. There is probably a migration of this species from the east to the west side of the island at the beginning of the N.E. monsoon, at which time no doubt many of these birds also come from India. A specimen obtained at Aripo in October is of a much richer brown than others I shot at Nuwara Eliya in February. These birds are fond of perching on the extreme top of a bush.

Blyth (Ibis, 1867, p. 304) refers the birds described by Layard to "*L. lucionensis,* Scopoli (?)," a race of *L. cristatus,* "distinguished by its prevalent ashy-brown hue." This character is not uncommon in Ceylon specimens which have old, worn plumage; but I have not seen it in newly moulted birds.

Bill dusky; irides dark brown; feet dark leaden.

Ceylon, India, Andamans, Malacca.

104. TEPHRODORNIS PONDICERIANA, Gmel.

A careful comparison of a series of *T. affinis*, Bl., from Ceylon, with a number of *T. pondiceriana* from India, has satisfied me that there is not sufficient ground for separating them specifically. The Ceylon birds appear to be smaller; but the depth of the general ashy brown of the upper surface varies in both, and to much the same extent. The supercilium also varies in distinctness in the birds from the two countries, but in those from India the maximum development is perhaps greater than in specimens from Ceylon.

These birds are common in the north and west of the island during the winter months, and probably migrate from the eastern side. They breed early in the year; and the young birds in their spotted plumage have been procured by Mr. Legge, in April, from the cinnamon-gardens at Colombo. Birds of the year are paler than in the following season. This may not have been known to Mr. Blyth when he described the Ceylon species as greyer than those from India.

Bill dusky; irides dull yellow; feet dusky lead-colour.

Ceylon, India, Assam, Upper Burmah.

105. HEMIPUS PICATUS, Sykes.

This bird is rare in the low country, and seems to be chiefly found on the upper hills. It is a common bird at Nuwara Eliya throughout the year, frequenting high bush jungle or low trees. Young birds have the colours less decided than adults.

Bill black; irides yellow; feet black.

Ceylon, S. India.

106. VOLVOCIVORA SYKESII, Strickl.

Generally distributed over the low country; it is resident in the Aripo district, and I have found it common near Colombo and in the extreme south. Although I have shot a great many of these birds, I have never obtained a female with any other than the barred under plumage and the grey head, and I cannot confirm Blyth's statement that the adult female has a black head and neck as in the male. The black in the young male first appears in spots on the top and sides of the head.

Bill black; irides brown; feet black.

Ceylon, India.

107. GRAUCALUS LAYARDI, Blyth.

Graucalus pusillus, Bl.

A smaller bird than the N. Indian *G. macei*, Less., with which it has been confounded. It differs also (Jerdon, 'Ibis,' 1872, p. 117) in having the under wing-coverts strongly barred, the abdominal bars absent in the adult male, and the outer tail-feathers only slightly tipped with white.

I have not seen this species alive; but it is occasionally found in the Kandy district and, according to Layard, who speaks of it as *G. macei*, in the S. and W. provinces.

Ceylon, S. India.

108. PERICROCOTUS FLAMMEUS, Forster.

Widely distributed, but nowhere very common. I have not met with it at Aripo, but have obtained it near Colombo; and it is tolerably numerous in the cold season on the hills. I have seen it more abundantly at Nuwara Eliya than elsewhere. It is generally in pairs, and perches high up on the trees.

Bill black; irides brown; feet black.

Ceylon, India, Assam.

109. PERICROCOTUS PEREGRINUS, Linn.

Common all over the island. It is resident in the Aripo district, and is found at Nuwara Eliya in the cold season. It does not frequent trees so much as bush-jungle; and I have never observed it perching very high, as is the marked habit of the preceding species.

Bill black; irides brown; feet black.

Ceylon, India, Andamans, Burmah.

110. BUCHANGA MINOR, Blyth.

This bird has been separated from the common Indian species (*B. macrocerca*), which it resembles in colouring, but from which it differs in all its dimensions. My finest specimen of *B. minor* is 10·5 inches total length instead of 12; wing 5·25 instead of 5·75 or 6; and other parts in proportion. The tail of the Ceylon bird is always less deeply forked than in the Indian species; and the small white rictal spot is frequently absent. Whatever may be thought of the value of these differences, they are constant; and I have not heard of the larger *B. macrocerca* of India being found in Ceylon. *B. minor* is abundant in the north; it is very common at Aripo, and is the only species of Drongo Shrike I have seen there. It is also found about Colombo, but by no means commonly within my experience. Its place there and in the south is occupied by another species. None of the Drongo Shrikes in Ceylon go above the lower hills, and for the most part they are confined to the low country.

Bill black; irides red-brown; feet black.

Ceylon.

111. BUCHANGA LONGICAUDATA, A. Hay.

I have seen this species on the tops of the trees in forest-jungle between Kandy and Trincomalie, and shot one specimen in a small wood about sixteen miles from Colombo. Layard says it is common in the Jaffna peninsula, and that "it frequents open lands, and perches on the backs of cattle to seek for ticks, on which it feeds largely." There must surely be some mistake about the species to which Layard here refers. His account agrees precisely with the habits of *B. minor*; and Lord Walden, who first described *B. longicaudata*, from India, where it is well known, tells me it is strictly a forest species, frequenting high trees, and is never seen on the backs of cattle.

Bill black ; irides red-brown ; feet black.

Ceylon, India.

112. BUCHANGA CÆRULESCENS, Linn.

Layard speaks of having procured one or two specimens of this species at Point Pedro, in the extreme north ; but it is not otherwise known from Ceylon.

Ceylon, India.

113. BUCHANGA LEUCOPYGIALIS, Blyth.

Peculiar to Ceylon. Allied to *B. cærulescens* ; but differing from that species in being smaller and having the dark grey of the breast continued (but gradually becoming paler) towards the vent, with the white confined to the under tail-coverts. In the immature bird the whole of the abdominal region is very dark grey, and the under tail-coverts have three or four broad dark bands on a paler ground.

This is the common species about Colombo and in the southern district. I have never seen it in the north.

Bill black ; irides brown ; feet black.

Ceylon.

114. DISSEMURUS LOPHORHINUS, Vieill.

Peculiar to Ceylon. *D. edoliiformis*, Blyth, is apparently the same species. The typical character consists in having the head subcrested, with the simple form of tail found in *Buchanga*. In Lord Walden's large collection of *Dicruri* from Ceylon, there are many examples showing an apparent gradation in the form of the tail between this species and *D. malabaricus* ; but as the true *D. lophorhinus* is found in localities where the racket-tailed species is unknown, I shall keep them distinct, and in my notice of the next species refer to the apparent gradations between them.

D. lophorhinus is found on some of the lower hills, and in wild districts in the low country in the southern half of the island. It appears to be quite a jungle bird.

Bill black ; irides brown ; feet black.

Ceylon.

115. DISSEMURUS MALABARICUS, Scop.

This is no doubt the species referred to by Layard under *Edolius paradiseus*, Linn., as it has been obtained in abundance in the district where Layard procured his specimens. It is quite confined to the jungle, and frequents the forests in the northern and central parts of the island. An immature specimen I shot in very wild country between Kandy and Trincomalie has the outer tail-feathers three inches longer than the next ; no part of the stem is bare ; but the inner web is very much narrowed just on a level with the tip of the adjoining feather. Lord Walden has received many similar specimens and others with the long racket-feathers in different stages of

growth, one example showing a difference in the intermediate length
of bare stem in the growing feathers on the two sides of the tail.
The short tail-feathers in some specimens appear to me to be possibly
a character of youth ; but they are regarded by Lord Walden as in-
dividual variations ; and the attention he has given to the *Dicruri* en-
titles his opinion to considerable weight.

Bill black ; irides brown ; feet black.

Ceylon, S. India.

116. ARTAMUS FUSCUS, Vieill.

Generally distributed over the low country, but is locally abundant
at certain seasons. It is very common at Aripo and in the neighbourhood of Colombo during the N.E. monsoon. I have always
found it in small parties and easy of approach.

Bill pale blue ; irides dark brown ; feet dark slate.

Ceylon, India, Burmah.

117. TCHITREA PARADISI, Linn.

Generally distributed over the low country, and, at certain seasons,
not uncommon on the lower hills. It is, however, a great wanderer,
and very uncertain in its movements. At Aripo they have at times
been very numerous ; and then I have not seen one for several weeks.
They seem to be of a fearless disposition, and used sometimes to fly
up under the roof of my veranda after spiders when I was standing
within a few yards of them. I have procured specimens in all states
of plumage at different times, and two examples showing the change
from the red to the white feathers—one of them at Aripo, the other
at Colombo, and both in January. Layard obtained one in February
in which the change was far advanced. They are common about
Kandy towards the end of the year.

Bill leaden blue ; irides brown ; feet pale blue.

Ceylon, India.

118. MYIAGRA AZUREA, Bodd.

Widely distributed, according to Layard, who, however, speaks of
this bird as *M. cærulea*, Vieill. I have only seen it from the western
province, where it is locally not uncommon.

Ceylon, India to China, Malaya, Andamans?, Philippines.

119. LEUCOCERCA AUREOLA, Less.

Leucocerca albofrontata, Frankl.

Layard records the occurrence of a *Leucocerca* in Ceylon which
was described by Blyth as *L. compressirostris*, from its differing from
the above species in having the bill more compressed. Mr. Blyth
tells me, however, that he believes now it was only a variety and that
it should come under the above heading. I have examined a Ceylon
specimen of *L. compressirostris*; and the character of the bill is very
decided, so much so as almost to justify the separation of the bird
from the *Myiagrinæ*, if in other respects it did not agree so closely
with *L. albofrontata*. That species, however, has been received from

Ceylon; and *L. compressirostris* may perhaps best be considered a
variety of it.

Ceylon, India.

Leucocerca ——?   Mr. Hugh Nevill (J. R. A. S., Cey. Br., 1867–
70, pt. i. p. 138) records the occurrence in the country round Nuwara
Eliya of a Flycatcher which he calls *Leucocerca fuscoventris*, Franklin.
The characters he gives of the species are evidently taken from a
description of *L. pectoralis*, Jerdon; and from what I know of the
circumstances I believe I am quite justified in saying that Mr. Nevill
never saw the Nuwara-Eliya bird except alive in the jungle.

It is not unlikely, however, that a *Leucocerca*, probably *L. pectoralis*, may be found on the Nuwara-Eliya hills, although it has not
yet been clearly identified.

120. MYIALESTES CINEREOCAPILLA, Vieill.

Resident and very common in the Nuwara Eliya district.   It frequents the lower branches of trees and is very bold and familiar, so
much so as to be rather a pest when one is collecting in the jungle,
from its habit of following one about or flitting from branch to
branch just in front of one.   I believe that in Ceylon it is almost
entirely confined to the upper hills.

Bill dusky; irides brown; feet fleshy brown.

Ceylon, India to Burmah and Tenasserim, China.

121. ALSEONAX LATIROSTRIS, Raffles.

A winter visitor to Ceylon, arriving early in October at Aripo.
It is common there until April, and is also found about Colombo at
the same season.   Its manners are precisely the same as those of the
European *B. grisola*, to which species it is closely allied.

Bill black, base yellow; irides brown; feet black.

Ceylon, S. India, China.

122. ALSEONAX TERRICOLOR, Hodg.

Butalis muttui, Layard, described by him from a single specimen
obtained at Point Pedro, agrees with Hodgson's *A. terricolor* from
India.   It is very rare in Ceylon.

Ceylon, North and Central India.

123. OCHROMELA NIGRORUFA, Jerdon.

Jerdon says of this Flycatcher that "it has hitherto (1862) only
been found on the summit of the Neilgherries and highest mountains
of Ceylon."   I can find no record, however, of this species occurring
in Ceylon except that by Layard, who says he saw a drawing made
by Mr. E. L. Mitford from a specimen he obtained at Ratnapoora.
This is in the low country and probably not a hundred feet above
the level of the sea.

124. EUMYIAS SORDIDA, Walden.

This species, distinguished and described by Lord Walden (Ann.
Nat. Hist. 1870, p. 218), is very common at Nuwara Eliya.   I have

observed it there at all seasons; but it appears to have been mistaken for *E. melanops*, Vigors, by both Kelaart and Layard, who evidently refer to it under that name in their catalogues. A specimen sent by Layard is in the British Museum, and is given in Gray's 'Hand-list' (4897) as *E. ceylonensis*, n. sp.?

The general colour of the head, back, and outer edges of the quill-feathers is a dark bluish grey; throat and breast more dingy, and becoming paler towards the vent; forehead and chin bright blue; wings and tail dusky.

Bill black; irides brown; feet black.

Ceylon.

125. CYORNIS RUBECULOIDES, Vigors.

Probably only an occasional visitor to Ceylon. Layard records having obtained a few specimens in the north of the island in October 1851; and I have examined specimens from Ceylon in Lord Walden's collection. Examples of this species from Ceylon and Burmah differ from Indian birds in having the orange colouring of the breast running up the centre of the throat, a peculiarity pointed out to me by Lord Walden.

Ceylon, India, Burmah.

126. CYORNIS JERDONI, G. R. Gray.

This species was at one time considered identical with *C. banyumas*, Horsf., from Java, and is given under that name by Jerdon in his 'Birds of India;' but it has been separated by Mr. G. R. Gray as distinct. It is a resident in Ceylon and not uncommon in the low country between Colombo and Kandy, but has not been recognized as being widely distributed. I have obtained specimens a few miles from Colombo in July. Mr. Legge describes the female as being brighter on the upper surface than the male, but this is not in accordance with what I have observed.

Bill black; irides brown; feet lavender (brown in dry skins).

S. India and Ceylon.

127. ERYTHROSTERNA HYPERYTHRA, Cabanis.   (Plate XVII.)

This Robin Flycatcher was described in 1866 by Cabanis (Journ. f. Orn. p. 391) from a specimen sent from Ceylon by my friend Mr. Nietner; and that example (in the Berlin Museum) was, I believe, the only one until now which had been brought to Europe. I was fortunate enough to obtain two specimens of this species at Nuwara Eliya in February 1870, and I have no doubt that it is not uncommon on the hills at that season. Mr. Nietner probably obtained his bird on his estate about 2000 feet below Nuwara Eliya; and further inquiries may perhaps lead to its discovery on the Neilgherries*.

---

* Since the above was written, a specimen of this Flycatcher has been sent home from Goona, Central India. It is in full breeding-plumage, and was supposed to be *E. parva*. It is very probable that these two species have been confounded when not in full plumage, and that *E. hyperythra* is not so rare or so local as appears to be the case at present.

The distinguishing characters of the species are the rich orange-brown of the throat and breast, and the black stripe running from the bill down the sides of the neck to the breast and terminating below the bend of the closed wing.   The specimens I obtained were both males, adult and immature; and the above characters are distinct in both, but much more so in the older bird.

These birds frequented low thin jungle; and I did not hear them utter any note.

Bill dusky above, yellow beneath; irides dark brown; feet purplish brown.

Ceylon, Central India.

Jerdon mentions that *E. leucura* is found in Ceylon; but I cannot find any special record of its occurrence there.   It may have been confounded with *E. hyperythra*.

128. Brachypteryx (?) palliseri, Blyth.   (Plate XVIII.)

Peculiar to Ceylon.   The generic position of this bird is not very clear.   It was placed by Blyth doubtfully in *Brachypteryx*, but differs from the birds of that genus in the sexes being alike in colouring and in the well-developed tail.   I believe it will require generic distinction; but for the present I shall leave it in *Brachypteryx*.

It is a species confined to the upper hills, and is by no means uncommon in the Nuwara Eliya district; but, from its habits, it is not an easy bird to watch or to obtain.   It frequents the low brushwood in the true jungle, creeping about the stems of the underwood close to the ground, and may sometimes be seen busily examining the dead branches of some fallen tree.   Frequently it betrays its close neighbourhood by its "cheep" once or twice repeated; and it will show itself for a moment within two or three yards of one; then it is lost again in the thick jungle.   By giving up a good deal of time I succeeded in obtaining a few specimens; but I have often been out for many hours without being able to get a shot, although I have occasionally heard the bird close to me.   It will sometimes show itself on a jungle-path; but it then keeps close to the side, turning over the dead leaves in search of insects, and disappearing on the slightest alarm.   When on the ground it often jerks its tail up after the manner of the Robins; but I have not observed this habit when it has been on the stems of the jungle plants or creeping about the dry sticks.   The sexes are alike in colouring.   I have one specimen which on dissection proved to be undoubtedly a male; and it could not be distinguished by any external character from the female.   Two other birds, of different sex and evidently young, were also alike, and differed from the adults only in the absence of the rusty throat and dark grey cheeks, and in having the tail shorter.   I have been unable to ascertain any thing of the nesting-habits of this species; and the bird itself is exceedingly rare in collections.

The whole upper surface is of a dark olive-brown, the wings, rump, and tail being of a richer brown tint; chin and throat pale rusty, beneath the eye and the ear-coverts dark greyish; the underparts pale olive, becoming brown at the flanks, vent, and under tail-coverts.

Bill dusky above, dark grey below ; irides pale buff ; fect dark flesh.

Ceylon (upper hills).

129. ARRENGA BLIGHI, n. sp.   (Plate XIX.)

In the adult, or perhaps nearly adult, male the whole head, nape, and throat pure black ; back, wing-coverts, and breast black strongly glossed with indigo ; carpal joint dark smalt-blue ; wings, tail, rump, flanks, and abdomen dusky brown, the two last slightly rufous. The upper tail-coverts, rump, and flanks are tinged with blue ; and it is not improbable that in an older bird these parts may become of the same colour as the back and breast. In the young the whole bird is brown, darker on the upper surface and more rufous below, the feathers of the forehead, throat, and breast centred with yellow-brown, and there is an indication of blue on the carpal joint.

The dimensions of the adult male are :—length 8 inches, wing 4·4, tail 3·5, tarsus 1·4, bill at front 0·6.

Bill black ; irides greyish ; feet black.

An adult female, shot by Mr. Bligh, but almost knocked to pieces, had very much the character of a young bird of the same sex I obtained at Nuwara Eliya (fig. 2) ; and the wing-spot was brighter, but not of so deep a blue as in the male.

The only example of this new species of *Myiophonus* I saw in Ceylon was the immature bird I obtained at Nuwara Eliya in July 1870 ; and the tinge of blue on the wing led Mr. Samuel Bligh of Ceylon to the opinion that it was the young of a species he had shot on the hills two or three years before, and which had been sent with other skins to Mr. Master of Norwich. By the kindness of that gentleman I have been able to examine his specimen and compare it with the one I myself obtained. There is no doubt of their belonging to the same species ; and as it has hitherto been unknown I have named it after my friend Mr. Bligh, who procured this first specimen of what is entirely a new form in the island.

Some credit is due to Mr. Edward Blyth for his remarks on the absence of certain birds on the Ceylon hills. He says ('Ibis,' 1867, p. 312) "That *Myiophonus horsfieldi* (or a specialized representative of this bird) has not been observed in the island is worthy of notice ; but I have before expressed an opinion that the higher regions of Ceylon have not yet been sufficiently explored." At the time Mr. Blyth wrote this the first specimen of the Ceylon *Myiophonus* was probably on its way to England ; and its true character has only now been recognized. Its nearest ally is *A. cyanea*, Horsf., from Java.

The habits of the Ceylon bird correspond, so far as is known, with those of the other *Myiophoni*. The young bird I procured at Nuwara Eliya was killed on a low branch of a jungle tree close to a little mountain-stream ; and Mr. Bligh, who obtained his specimens at an elevation of between 4000 and 5000 feet, told me he had never met with the bird excepting in the immediate neighbourhood of water-courses. He writes me that although he has seen this species several

times it is very difficult to obtain. The bird frequently perches on
a rock in the midst of some mountain-torrent, but is very impatient
of observation. On these occasions it "gives utterance to a pecu-
liarly long-drawn, plaintive though loud whistling note; at the same
time the body is dipped and the tail slightly raised." It soon seeks
shelter under the deuse jungle foliage.

130. Pitta brachyura, Linn.

Generally distributed in Ceylon during the winter months, and at
that time very abundant at Aripo. Although most of these birds
seen in Ceylon are probably visitors from India coming in October,
I have reason to think some of them are residents, as I have frequently
heard and more than once seen them at Nuwara Eliya in August.
My house at Aripo was surrounded by *Suriya* trees, the branches of
many of them touching the roof of the veranda; and to these trees
the Pittas used to come every evening shortly before sunset, perching
about six or eight feet from the ground and continually repeating
their cry of "A-vitch-i-a" (the name given to the bird by the Singhalese), which was frequently followed by a low hissing scream. On
being alarmed by my too close approach they would fly direct to the
hedge about thirty yards distant and hide themselves under the
darkest and thickest part of it. A frequent attitude of this bird when
perched on a stout branch of a tree was with the head and body
stretched up to the full height, the legs straight, and the tail turned
upwards.

Bill orange, tip dusky; irides brown; feet flesh-colour.
Ceylon, India.

131. Geocichla layardi, Walden.

Peculiar to Ceylon. A single example of this Thrush was sent to
Lord Walden in a collection of birds from the island. From what I
heard in Ceylon from the person who made the collection I have no
doubt this bird was obtained on the hills on the south-east side of the
island, a part of the country which has not yet been properly examined
and is likely to produce more novelties. This bird is described as
more nearly allied to *G. citrina* of North and Central India than to
G. cyanota of Malabar, with the orange colour of the underparts
brighter and richer than in *G. citrina*, but not nearly so deep as in
G. rubecula of Java.

The colours of this specimen are rich orange on the head, neck,
and underparts, bluish grey above, and a white spot on the wing.
Ceylon.

132. Turdulus wardii, Jerdon.

Generally a rare bird in Ceylon; but Mr. Laurie tells me it is not
uncommon during the north-east monsoon in some of the hill-forests.
I have seen specimens collected by that gentleman and others from
the Kandy district, but have not met with the bird alive.
Ceylon, India.

133. MERULA KINNISI (Kelaart), Blyth.

Peculiar to Ceylon, and, I believe, confined to the upper hills. It is very common at Nuwara Eliya, frequenting alike the edges of the jungle and the gardens of the English houses, and often building in the stables and outhouses. It has the habits generally of the English Blackbird; but its song is by no means so fine.

The male has the whole upper surface black with a bluish-grey tinge, the underparts more dingy; the female has the upper colour less intense, and is dark ashy brown below. Young birds have the head and back brown, with the throat and breast mottled, the feathers being pale-centred and with dark brown tips.

Bill bright orange (adult), yellowish brown (young); irides brown; orbits yellow; feet yellow.

Ceylon.

134. OREOCINCLA NILGIRIENSIS, Blyth.

This handsome long-billed Thrush was described by Layard under the name of *Zoothera imbricata* from a specimen received from Mr. Thwaites, who probably obtained it on the hills. It has since been recognized as the above species. I have examined two skins sent home by Mr. Bligh, and Layard's specimen now in the British Museum; and the scale-like appearance of their plumage, arising from the black border to each feather, is well marked.

"Bill corneous; legs brown" (*Layard*).

Ceylon; Neilgherries.

135. OREOCINCLA SPILOPTERA, Blyth.

Peculiar to Ceylon. This is quite a jungle bird and not very uncommon in suitable places on the hills. Many specimens have been procured in wild country not far from Kandy, and in the forest-land adjoining the coffee estates between 2000 and 5000 feet high. I have not met with it at Nuwara Eliya.

Bill black; irides brown; feet pale brown.

Ceylon.

136. PYCTORHIS SINENSIS, Gmel.

Layard observed this bird in widely separated localities in the low country, but does not speak of it as numerous. I have seen a specimen in the possession of Mr. Legge, R.A., at Colombo, which I believe he told me was killed near his house; and I have seen others from the Kandy country.

Ceylon, India to Burmah and China?

137. ALCIPPE NIGRIFRONS, Blyth.

Peculiar to Ceylon. This little bird is well distinguished from the allied species *A. atriceps* by the greater part of the head being brown, the black being confined to the forehead, and a broad streak through the eye to the ear instead of covering the whole top of the head. I have not seen this bird in the north of Ceylon; and Layard does not say where he discovered it; but it is abundant in the central and probably in the southern districts.   It is, however, somewhat migratory within the island, and it is difficult to say to what cause its irregular movements are due.   I have shot it both near Colombo and at Nuwara Eliya in January, and have found it abundant at the latter place in July, August, and September; then it has entirely disappeared.   It is an amusing little bird, usually found in small parties and frequenting underwood and low thick bushes, or creeping among the stems of the taller jungle-plants, occasionally coming to the edge of a path and betraying its presence by an angry hissing note, evidently intended to warn off intruders.

Bill dusky above, pale flesh-colour beneath; irides golden; feet purplish flesh.

Ceylon.

138. DUMETIA ALBOGULARIS, Blyth.

This species is said by Layard to be confined to the vicinity of Colombo; and although it is unlikely to be so purely local, I certainly never saw the bird alive until I became a resident close to the cinnamon-gardens in which he observed it.   Like the following species it will probably be found in bush jungle in the interior as well as in the immediate neighbourhood of Colombo.

Ceylon, S. India.

139. DRYMOCATAPHUS FUSCICAPILLUS, Blyth.

Peculiar to Ceylon and rarely met with.   I only know of three* specimens having been obtained—two of them by Layard in Colombo and in the central road leading from Kandy northwards, and one (a male) by myself also from the latter part of the island.   I found this bird among thick underwood in forest-jungle by the side of the road on which I was travelling; and it was perched within two feet of the ground when I had my first fair view of it as with outstretched neck and swelling throat it poured forth a torrent of babbling notes.

I have restored this bird to its original position in the genus *Drymocataphus*, as its bill does not agree in form with that of *Pellorneum*, and the fifth quill-feather is the longest, the fourth and sixth being equal and slightly shorter.   The colour of the back, wings, and tail dark olive-brown, the last tipped rufous; wings and tail in some lights showing distinct transverse striæ; crown rich dark brown, the feathers slightly pale-shafted; lores, cheeks, sides of neck, and all the underparts pale rufous brown, the breast being rather darker.

Bill dusky above, flesh-colour below; irides red; orbits yellow; feet pale flesh.

Ceylon.

140. POMATORHINUS MELANURUS, Blyth.

Peculiar to Ceylon; rather local in its distribution, but generally numerous where it is found.   It is very abundant at all times of the

* I have since seen a fourth, which was procured a few years ago by Mr. Bligh from the hills.

year at Nuwara Eliya and in the surrounding district, frequenting the primitive jungle with which the upper hills are covered. It is also found occasionally in wild country near Kandy, and was first seen by Layard "in low, scrubby, and almost impenetrable brushwood" a few miles from Colombo. It was probably not far from this last locality that I also met with it, in the low country, a wild district of no great extent, to which I have referred in my notice of *Harpactes fasciatus*. Like its congeners, however, this Scimitar-bill is essentially a hill bird. It creeps about underwood and the lower branches of trees, half opening and closing its wings, and assuming various kinds of strange attitudes. It is at all seasons noisy ; and just about the pairing-time in February the cries of a party of these birds remind one more of a concert of Cats than any thing else. It is to this species the name of Gamut-bird is often applied, from the powerful notes of the male beginning very low and running up the scale ; they have a very striking sound when heard amid the silence of the deep jungle.

The colour of the sexes is alike. The back, wings, flanks, vent, and under tail-coverts rich olive-brown with a rufous tinge, especially on the flanks ; from the base of the upper mandible to the nape black, extending to the mixed olive-brown and black on the top of the head : throat, breast, middle of abdomen, and a conspicuous supercilium pure silky white ; tail blackish brown. The young bird is much more rufous generally, and has the ear-coverts and the sides of the neck and breast quite rusty.

Lord Walden has a series of specimens of *Pomatorhinus* the localities of which are not very intelligible on the labels ; but the birds were probably obtained in the south or south-east of the island. All these have the upper surface quite rufous, extending also to the tail. This colouring is not found in one of the many specimens I have from Nuwara Eliya, and is so marked as almost to justify a specific distinction.

Bill yellow, with the base dusky above ; irides dark red ; feet lead-colour.

Ceylon.

141. GARRULAX CINEREIFRONS, Blyth.

Peculiar to Ceylon. This species is confined to the southern half of the island, frequenting the lower hills, and, according to Layard, "it much resembles the *Malacocerci*, hunting in small parties and incessantly calling to each other." It is not uncommon in the Kandy district and in the hilly country between that and Galle. I have examined a great number of specimens of this species, and have found them agree very closely with each other ; but they differ so materially in dimensions from those given by Blyth that I can only suppose he had but one example before him, and that an immature bird. This impression is confirmed by the specific name *cinereifrons*, given by him, and agreeing with his description "forehead and cheeks pale ashy ;" whereas the birds I have examined have the whole top of the head ashy, that colour often extending over the nape, as well as the cheeks, which are paler than the rest of the head ; chin albescent, becoming rufous on the throat; in other respects the colours agree with Blyth's description. The dimensions of a specimen I obtained at Kandy, and which is not at all unnaturally stretched out, but fairly represents an adult bird, measures fully 10 inches instead of 8·5 ; the other comparative dimensions are :—wing 4·75, 4·5 ; tail 4·5, 4 ; bill to gape 1·3, 1·25 ; tarsus 1·5, 1·25.

Bill black ; irides buff; feet dusky.

Ceylon.

142. MALACOCERCUS STRIATUS, Swains.

A comparison of specimens of *M. striatus* I obtained in Ceylon with *M. malabaricus* in the Calcutta Museum left me in great doubt as to the reason for separating them specifically, and I cannot but think they will ultimately be included under the same name. The depth of the striæ in *M. striatus* varies with age ; in a well-grown young bird there is not a trace of striæ on the tertiaries, and they are very indistinct on the tail. In a fully adult bird now before me the striation exactly agrees with Jerdon's description of that character in *M. malabaricus*: "the tertiaries are but very obscurely striated, but the tail is distinctly so." The distinctive character of *M. striatus* has hitherto been shown by comparing it with *M. terricolor*; but it should have been placed by the side of the Malabar species.

The Ceylon bird is universally distributed over the low country, frequenting alike the jungle, half-cultivated ground, and the gardens and compounds in Colombo. Its manners are the same as those of the common Indian species. I have found it nesting at Aripo in January.

Young birds are slightly rufous.

Bill pale yellow ; irides pale buff; feet pale yellow.

Ceylon, S. India?

The only record I can find of the occurrence of *M. griseus*, Gmel., in Ceylon is in the 'Appendix' to Kelaart's 'Prodromus Faunæ Zeylanicæ' (p. 45), where, in a report by Mr. Blyth on a collection of Ceylon Mammals, Birds, Reptiles, and Fishes, and, I presume, made to the Asiatic Society of Bengal, the following appears among the list of birds :—

"*Malacocercus griseus* (Lath.), var.—Resembling the species of S. India, except that the head is concolorous with the rest of the upper parts."

I have neither seen nor heard of the true *M. griseus* in Ceylon.

143. LAYARDIA RUFESCENS, Blyth.

Peculiar to Ceylon, and tolerably common in the wilder parts of the low country in the southern half of the island. It was formerly considered a hill species ; but I believe it only visits the upper hills during the cold season. I have only found it at Nuwara Eliya at the beginning of the year; but it is at all times to be met with a few miles from Colombo where there is jungly or half-cultivated land.

It keeps in small parties, and has generally the habits of the other *Malacocerci.*
Bill dull orange ; irides white ; feet yellow.
Ceylon.

144. HYPSIPETES GANEESA, Sykes.

As *H. neilgherriensis,* Jerdon, is now united with *H. ganeesa,* Sykes, the Ceylon birds will come under the latter title.
This species in Ceylon is, I believe, confined to the hills, and is most abundant at a moderate elevation. I have only seen it at Nuwara Eliya in February ; but it is tolerably common in jungle from the Kandy country to about 5000 feet. I have generally found it in small parties on rather low trees.
Ceylon, S. India.

145. CRINIGER ICTERICUS, Strickland.

Layard says of this bird that it "abounds in the mountain zone." This probably means the lower ranges, as he tells me he has never visited the Nuwara-Eliya district, and he does not profess to know the hill birds. Kelaart, on the other hand, who specially collected the birds of the upper hills, says it is "a common species in the low country." I have no doubt Layard was right in suggesting that Kelaart mistook the common *Ixos luteolus* for this species ; and this is confirmed by Mr. Legge's observation that *Criniger ictericus* "is strictly a jungle bird" (J. R. A. S., Ceylon Branch, 1870–71, p. 43). Mr. Legge, however, whose knowledge of the low country at the time he wrote was confined to the western province, says "Kelaart wrote correctly of this bird ;" but "a strictly jungle bird" can hardly be described as common in a district principally consisting of paddy-fields and cultivated land.
I have only obtained this bird once in the neighbourhood of Colombo, among trees in a native village ; it is most numerous in forest country on the lower hills, as is the case with this species in India.
Bill black ; irides red ; feet dark leaden.
Ceylon, Malabar.

146. IXOS LUTEOLUS, Less.

This bird, the *Pycnonotus flavirictus* of Strickland, is one of the commonest species in the low country. It is equally abundant at Aripo and Colombo wherever there are low bushes, and has a hurried twittering song of a few notes, loud and frequently repeated.
Bill black ; irides red ; feet blue-black.
Ceylon, South and Central India.

147. KELAARTIA PENICILLATA, Blyth.

Very abundant at Nuwara Eliya and on the upper hills, frequenting low bushes and thin jungle.
The general colour of this bird is dark olive-green above and greenish yellow below, brighter yellow on the throat, middle of

abdomen, and under tail-coverts; head black in front aud shading
into olive-green at the nape, with the feathers tipped paler or white;
chin and a narrow vertical stripe on each side of the forehead white;
lores and cheeks black, paling to leaden grey behind, with a yellow
spot below the ear and a tuft of bright yellow feathers springing from
immediately behind the eye and directed backwards. These tufts
stand out from the sides of the head when the bird is alive, and add
much to its generally handsome appearance.

This species was described from Ceylon specimens; but is believed
by Jerdon to be "identical with one procured from the Mysore country
below the Neilgherries, which was accidentally destroyed," but from
which a coloured sketch was made.

Bill black; irides red-brown; feet leaden.

Ceylon, S. India?

148. RUBIGULA MELANICTERA, Gmel.

Peculiar to Ceylon, and tolerably common in the low country and
lower hills of the central and southern portions of the island. I have
obtained specimens near Colombo and close to Kandy.

The colour of the upper surface is olive-brown, and of the under
parts bright yellow, with the flanks tinged with dull olive; top and
sides of the head black; quills brownish black, with the outer edge
olive, and tail dingy black, with all but the central feathers tipped
white.

Bill black; irides red; feet purplish black.

Ceylon.

149. PYCNONOTUS HÆMORRHOUS, Gmel.

Very common all over the low country, and less so on the lower
hills.   I have never seen it at Nuwara Eliya or above 5000 feet, and
I am inclined to think it is only a seasonal visitor to that elevation.
I have found this species breeding in December at Aripo; its cup-
shaped nest was placed under the eaves of my bath-house out of doors,
and supported by the sticks of which the rough framework was con-
structed.

Bill black; irides dark brown; feet leaden black.

Ceylon, South and Central India.

150. PHYLLORNIS JERDONI, Blyth.

Common in the low country.   I have obtained it on several occasions at Aripo, near Colombo, and quite in the south.   It generally
keeps among the upper branches of the trees.

Bill slate; irides brown; feet lavender.

Ceylon, India.

151. PHYLLORNIS MALABARICUS, Lath.

Recorded by Layard and Kelaart from the hills; and I have seen
a specimen obtained by Mr. Laurie. I believe it is rather rare in
Ceylon.

Ceylon, South and Central India.

152. IORA ZEYLONICA, Gmel.

This species is well known in the low country; it is very abundant at Aripo at all seasons, and almost as common about Colombo. It breeds at Aripo at the end of the year; and I have obtained it in November with hardly a trace of green on the black back. Its notes are very much varied; and some of them sound as if uttered at a considerable distance when the bird is really within a few yards.

I regret that I did not know whilst I was in Ceylon of the question as to *I. typhia*, Linn., being found in the south of India and Ceylon. As I brought home no male specimens of *Iora* which were not in such a state of plumage as to leave a doubt about their belonging to *I. zeylonica*, I shall not include *I. typhia* among the Ceylon species; but I have a very strong impression, partly based on my recollection of a pair of birds with dull green backs which for several days frequented some shrubs close to a house where I was staying, a few miles from Colombo, that *I. typhia* is found in Ceylon. I had no doubt of it at the time, as the male of *I. zeylonica* should then (February), according to my observations, have the back nearly or entirely black.

In case this paper should fall into the hands of any one collecting in Ceylon, but who is not familiar with the distinctive characters of the two species of *Iora*, I may mention that the females are at all times practically alike. In the breeding-season the male of *I. zeylonica* has the back entirely black or, more frequently, black and green irregularly mixed, the colours being in patches and not generally blending with each other; at the same season the male of *I. typhia* has the back wholly green, contrasting with the black wings, which in both species have two white bars. A further distinction is said to exist in the colour of the irides (this would hold good at all seasons), those of *I. zeylonica* being grey and those of *I. typhia* light hazel; I can answer for the former being correct.

Bill slate; irides grey; feet dull leaden.

Ceylon, S. India.

153. IRENA PUELLA, Latham.

Layard and Kelaart have each recorded an example of this species, both from near Kandy.

Specimens of this bird from Ceylon are much desired for comparison with those from India. The male has the whole upper parts and under tail-coverts bright cobalt-blue; wings, tail, and lower plumage deep velvet-black. The female is of a dull, slightly mottled Antwerp blue throughout. (*Jerdon.*)

Ceylon, Malabar, Assam, Arracan, Burmah.

154. ORIOLUS INDICUS, Jerd.

I include this species of Oriole on the authority of Layard, who speaks of a pair of these birds having been shot near Colombo, and coming under his notice.

Ceylon, India.

155. ORIOLUS CEYLONENSIS, Bouaparte.

Generally distributed in the low country. I have met with it commonly at Aripo, Colombo, and in the south ; but I have no reason to think it ascends above the lower hills. The young bird has the back pale dirty yellow, purer on the rump ; top of the head brownish black, becoming streaked on the cheeks and strongly so on the throat and under neck; quills margined externally with whitish, and the colours generally very much less pure than in the adult. The bill in the young is black.

Bill deep flesh-colour ; irides red ; feet leaden.

Ceylon, S. India.

156. COPSYCHUS SAULARIS, Linn.

Abundant in the low country, and rarely found far from native villages or the houses of English residents. The familiarity of the "Magpie Robin" makes it a general favourite ; and whether when perching on the roof of the house (a frequent station for it when singing) or furiously attacking some intruding rival, there is always something attractive in this showy and well-known species. During the last hour before sunset these birds become very noisy and frequent fights take place between the cocks, two or three of them going through a sort of tournament before the hen bird which has taken up her quarters in the neighbourhood. It is at this time the cocks put themselves in such strange attitudes, turning back the tail till it almost touches the head, as Layard mentions ; but Jerdon says he has never observed these performances, which from my own observation I should say are regularly gone through every afternoon ; the birds frequently utter a harsh kind of scream ; and this goes on until the sun disappears and the quickly following darkness puts an end to the proceedings.

Good specimens of this Robin are very difficult to obtain at Colombo, unless immediately after moulting ; as the birds soon become discoloured with the red soil, and the tails rapidly worn out at the end.

Females of this species from Ceylon have the back darker than those from Burmah, and perhaps from India generally, but they do not differ from a Madras specimen in the British Museum.

The young birds are greyish brown above, with the throat and breast mottled with dark brown on a paler ground, and the bill dusky.

Bill black ; irides brown ; feet dark leaden.

Ceylon, India, Arracan, Tenasserim, S. China, Hainan.

157. KITTACINCLA MACRURA, Gmel.

This bird is confined to wild jungly districts in the low country and on the lower hills. In such localities it is numerous and its fine song may be constantly heard in the morning and evening. It is abundant in the wilder parts of the northern road from Kandy ; and I have also heard it occasionally in a piece of thick jungle close to

Kandy itself. It usually perches low ; and from its habit of frequenting dense jungle, it is often difficult to obtain sight of.
Bill black ; irides brown ; feet flesh-colour.
Ceylon, India, Assam, Burmah, Malacca, Hainan.

158. THAMNOBIA FULICATA, Linn.

Common about houses and outbuildings, and, I believe, generally distributed through the lower parts of the country. I have seen them more numerous in the north than elsewhere ; and they were always about my house at Aripo, frequently coming into the veranda, and generally very tame.
Bill black ; irides brown ; feet black.
Ceylon, S. India.

159. PRATINCOLA CAPRATA, Linn.

Layard and Kelaart both mention having obtained this species on the lower hills ; but I have never met with it, either alive or as a skin.
Ceylon, India, Burmah, Malaya, Philippines.

160. PRATINCOLA ATRATA, Blyth.

Very common at Nuwara Eliya and on the upper hills. It frequents gardens rather than jungle ; and the top of a rhododendron bush is a favourite station for the male, which always chooses a conspicuous position when it sings its short Robin-like song. Young males at first have the general brown plumage and rufous rump of the female ; the change to the pure black and white of the adult male is very gradual, the quills and rump being the last to assume the mature colours.
Bill black ; irides brown ; feet black.
Ceylon hills, Neilgherries.

161. LARVIVORA CYANA, Hodgson.

This bird is a winter visitor to Ceylon. Layard obtained specimens in October in the extreme north ; and I procured adult and immature examples of both sexes at Nuwara Eliya in January, February, and March. It was at that time tolerably common on the hills ; but I have not met with it at any other season.
The female is olive-brown above ; underparts rufous, paler on the throat and centre of abdomen ; under tail-coverts white. These particulars, taken from one of my Ceylon specimens, agree with Hodgson's last description of the colours in the female.
Bill dusky ; irides brown ; feet flesh-colour.
Ceylon, India (generally on the hills).

162. CYANECULA SUECICA, Linn.

Layard obtained this species in March in one of the coffee-districts. I have not met with it.
Ceylon, India, N. and W. Asia, N. Europe.

163. ACROCEPHALUS DUMETORUM, Blyth.

This is apparently the bird given by Layard as *Phyllopneuste montanus*, Blyth.

Generally distributed; it is a winter visitor and numerous in Ceylon at that season. I have killed it at Aripo, Colombo, and Nuwara Eliya. All my specimens have the greenish shade on the upper surface mentioned by Blyth as found in the birds from Ceylon.

Bill dusky above, pale flesh below; irides brownish yellow; feet in different specimens pale brown to purplish flesh.

Ceylon, India, Nepal, Assam.

164. ORTHOTOMUS LONGICAUDA, Gmel.

Common in all parts of the island, but especially frequenting gardens and the neighbourhood of habitations. It is as abundant at Nuwara Eliya as at Aripo or other parts of the low country. I have examined many of these birds from different localities, and have found them to agree in all respects with Jerdon's description of this species, except in the length of the tail; this in Ceylon birds I have never found to exceed 2¼ inches.

Bill dusky flesh; irides yellow; feet flesh.

Ceylon, India to Burmah, S. China.

165. PRINIA SOCIALIS, Sykes.

Layard found this species in the extreme north; and I believe Mr. Legge discovered it nesting in a patch of Guinea grass close to his house at Colombo. It will probably be found in suitable situations in other parts of the island.

Ceylon, S. India.

166. CISTICOLA SCHŒNICOLA, Bonap.

Cisticola homalura, Blyth?

I place these two species together as it is difficult to speak of them separately, in consequence of the confusion existing between them, if they are really distinct. Layard says of "*C. cursitans*, Blyth" (? = *C. schœnicola*, Bonap.) that it "is much less common than *C. homalura*; and though found in the same locality, it frequents trees and jungle." This bird surely cannot be a *Cisticola*. Kelaart says of *C. cursitans* :—"frequents the grass-plains; very common at Trincomalie." I can confirm this statement and say precisely the same of it at Colombo; it is common there wherever there is a patch of long grass.

C. homalura was discovered by Layard in paddy-fields near Galle; he "subsequently found it sparingly about Colombo, and abundantly in fields of gingelle (*Sesamum orientale*) at Pt. Pedro." Kelaart says it "is found in great abundance on Horton Plains and Nuwera Ellia," these last localities resembling each other in being elevated grass-plains surrounded by forest-jungle.

I am almost ashamed to think of the number of specimens of *Cisticola* I have shot at Nuwara Eliya in the hope of getting one of

C. homalura, which Blyth says ('Ibis,' 1867, p. 302) "differs from *C. schœnicola* in having a stouter bill, the whole upper parts much darker, and the tail almost even, except that its outermost feathers are '25 inch shorter than the next;" but, except in some considerable variation occasionally in the depth of the general rufous tint, there was nothing to distinguish them from the grass-frequenting species at Colombo.   Mr. Layard tells me that the fine collection of Ceylon birds he brought to England is now in such a state as to be useless for scientific purposes; and as I can obtain no specimens of *C. homalura* for examination, I must regard that species as very doubtful until further evidence is procured from the localities whence Layard obtained his birds.

C. schœnicola from Ceylon agrees with the European bird in size, and is larger than the Indian representative; it has, however, the same decided markings as the latter form, and they are even more conspicuous.   The dimensions given by Jerdon are greater than those of any of the Indian specimens I have examined.

Bill dusky above, flesh below; irides pale yellow; feet flesh-colour.

Ceylon, India to Europe, Africa, China, Hainan, Formosa.

167. DRYMOIPUS INORNATUS, Sykes.

The difficulty in determining the species of *Drymoipus* is so well known that it may prevent additional confusion if I mention that the three species included in this list of Ceylon birds have been compared with specimens in the British Museum, and satisfactorily identified with the species there labelled with the names I have given.   The identification of at least one of the two Colombo species by myself and Mr. Legge whilst I was in Ceylon was not correct; and it is uncertain to which of them Mr. Legge's observations (J. R. A. S., C. B., 1870–71, p. 50) refer.

I believe *D. inornatus* is not uncommon about Colombo; but the only specimen I brought to England came from Kandy, and agreed with those in the British Museum in having the lores, throat, and cheeks whitish, the whole under surface and flanks very light, with a dull yellowish tinge, and a rather broad subterminal dusky band of uniform tint on the under surface of the tail-feathers.   The bill is rather slight and black, with the base of the under mandible abruptly pale (dried skin).   The wing exactly 2 inches.   Jerdon says of this species, "in no case does the wing ever come up to 2 inches, more generally 1¾." I cannot think, however, there is any doubt about this specimen being *D. inornatus*.   Layard says the eggs of this species are "verditer, with purplish blotches and wavy lines;" Mr. Legge gives "ground-colour clear blue-green, clouded here and there, or blotched mostly towards the obtuse end, with sepia." It is doubtful to which species either of these gentlemen refers.

168. DRYMOIPUS JERDONI, Blyth.

The common Ceylon species, of which I have obtained specimens close to Colombo, agrees perfectly with *D. jerdoni*, Blyth, in

the British Museum, where there is a specimen named and sent by
Dr. Jerdon himself. In the 'Birds of India,' vol. ii. p. 180, Jerdon
mentions that Blyth described this species from specimens he sent
him from Southern India; but he afterwards absorbed it into *D. lon-
gicaudatus* in the belief that the specimen he described was in imperfect plumage. Jerdon further says:—"It appears to me very
similar to some Ceylon birds which Mr. Blyth doubtfully considered
identical with *D. inornatus.*"

My Ceylon birds are greyish brown on the upper surface, rather
paler on the head, cheeks, and neck; lores pale and much less conspicuous than in *D. inornatus*; under surface pale fulvous, and flanks
rather dusky; the upper surface of the tail-feathers distinctly striated,
the striæ showing as faint narrow bars on the under surface, which
has a narrow dark subterminal band, generally darker in the centre,
and giving the appearance of a spot. In fresh specimens the bill is
dusky above, fleshy below; irides pale yellow; feet flesh-colour.
Length 5·5 inches, wing 2·3, tail 2·5, tarsus ·8, bill at front ·4.
Ceylon, S. India.

169. DRYMOIPUS VALIDUS, Blyth.

This species, at first called *D. robustus* by Blyth, is peculiar to Ceylon, and, according to Layard (who discovered it), rather a rare bird.
Mr. Legge and I were both mistaken in believing it common about
Colombo, as I now find I did not see the species in Ceylon. A specimen in Lord Walden's collection, agreeing with another in the
British Museum, has the bill entirely black, stouter and considerably
deeper than I have seen in any other Ceylon species; top of the
head, lores, and general upper surface dark greyish brown; beneath
whitish, with a pale fulvous tinge; cheeks, sides of the breast, and
flanks dusky. Length 6 inches, wing 2·4, tarsus 1, bill at front ·5.

The dry specimen has the bill black; tarsus yellow-brown (probably flesh-colour when alive); irides "light red-brown" (*Layard*).
Ceylon.

170. PHYLLOSCOPUS NITIDUS, Lath.

This bird is common at Nuwara Eliya in the cold season; and I
have seen it also at Aripo.
Bill dusky above, flesh below; irides dark brown; feet pale brown.
Ceylon, India.

171. PHYLLOSCOPUS VIRIDANUS, Blyth.

Recorded by Layard, who also gives *Phyllopneuste montanus*,
Blyth, which is probably a synonym of *Acrocephalus dumetorum*,
Blyth.

172. SYLVIA AFFINIS, Blyth.

I obtained one specimen of this species at Aripo in December.
Layard also appears to have only met with it on one occasion.
Bill, base slate, tip dusky; irides pale yellow; feet dark leaden.
In this species and very many others Jerdon has apparently given

the colour of the bill and legs from their appearance in dried specimens.
Ceylon, Central India.

173. MOTACILLA MADERASPATENSIS, Gmel.

Layard mentions having seen one specimen in a private collection in Ceylon.
Ceylon, India.

174. CALOBATES SULPHUREA, Bechst.

I have obtained this bird at Nuwara Eliya in the beginning of the cold season ; it is better known on the hills than in the low country.
Asia to Australia, Africa, Europe.

175. BUDYTES VIRIDIS, Gmel.

This is the common Wagtail in Ceylon, appearing with other migratory birds in October.
Ceylon, India, "N. Africa, S.E. Europe, and W. Asia" (Jerdon).

176. LIMONIDROMUS INDICUS, Gmel.

I have only seen this bird on wild jungle-roads between Kandy and Trincomalie ; but Layard has apparently met with it in other localities.
Ceylon, India, Arracan, Burmah, and part of Malaya, China.

177. CORYDALLA RICHARDI, Vieill.

This species is numerous in winter on the "Galle face"—the esplanade at Colombo, and a great place of resort for Pipits, Wagtails, and small Sand-Plovers at that season. It is no doubt, as Layard states, widely distributed ; but I do not think it is a resident in Ceylon.
Ceylon, India, and Asia generally, Africa, Europe.

178. CORYDALLA RUFULA, Vieill.

Resident and very common in Ceylon ; I have found it at Aripo, Colombo, and Nuwara Eliya ; and I believe it is generally distributed throughout the island.
Bill dark brown above, yellowish below ; irides brown ; feet light fleshy brown.
Ceylon, India, Assam, Burmah.

179. CORYDALLA STRIOLATA, Blyth.

This bird is also common at Colombo in the winter. I have compared and identified specimens of this and C. richardi from Ceylon with birds in the Indian Museum at Calcutta.
Ceylon, India, China.

180. ZOSTEROPS PALPEBROSUS, Temm. (Plate XX. fig. 1.)

Common in the central and southern parts of Ceylon, but only ascending the hills to about 2000 feet. It frequents trees and

flowering shrubs, and, Mr. Legge says, is often to be seen on the
tulip-trees in the principal street of the Fort at Colombo.   It is
common about Kandy and the surrounding district ; but I have never
met with it in the north or on the upper hills.   Specimens of this
Zosterops from the low country in Ceylon vary somewhat in size, but
have been identified in England and Calcutta with *Z. palpebrosus*,
and agree with Jerdon's description of that species except in being
generally smaller and in the colour of the bill and legs.   He says,
" Bill blackish, horny at the base beneath ; legs reddish horny ; "
but I find in freshly-killed birds the following colours :—
Bill dark leaden, paler at the base beneath ; irides light brown ;
legs and feet lavender.
Ceylon, India, Assam, Arracan, Tenasserim.

181. ZOSTEROPS CEYLONENSIS, n. sp.   (Plate XX. fig. 2.)

Upper surface dark olive-green, deeper on the head and paler on
the upper tail-coverts ; a circle of small white feathers round the eye ;
lores and below the eye dusky, but not very conspicuous ; chin, throat,
and centre of breast greenish yellow, shading at the sides of the neck
and breast into the colour of the back, and giving the appearance of
an incomplete pectoral band ; the rest of the underparts bluish white,
darkest on the flanks, and sometimes tinged in the centre with yellow ;
under tail-coverts yellow ; quills and tail dusky brown, both margined
externally with olive-green, and the latter faintly marked with transverse striæ.   Sexes alike.
Length 4·75 inches, wing 2·4, tail 1·8, bill at front ·5, tarsus ·7.
Bill dark leaden above, paler below ; irides light brown ; feet
lavender.
This is at all seasons one of the commonest birds at Nuwara Eliya
and on the upper hills.   It is, I have no doubt, the one recorded by
Kelaart as *Z. annulosus*, Swainson, an African species.   Layard, in
speaking of this bird in his ' Notes on the Ornithology of Ceylon,'
says : "Dr. Kelaart writes, 'we fear that the Nuwara Eliya *Zosterops*
is wrongly identified ; it is of a darker green than the common
Z. palpebrosus.'"   He then adds, " I, however, much doubt the distinctness of this and the preceding species."   A comparison of the
two birds, however, leaves no doubt that there is a marked difference
between them, both in colour and in the form of the bill.   The
bird from the Ceylon hills cannot be identified with any recognized species ; and Mr. A. O. Hume, to whom I showed specimens of
it when I was at Calcutta, told me he had never seen it in any of his
many collections from the Neilgherries, a district (as I have before
mentioned) agreeing closely in character and productions with the
Ceylon hills.   Mr. W. T. Blanford, in a paper on the Birds of
Western India (J. A. S. B. 1869, vol. xxxviii. p. 170), says, in speaking of *Z. palpebrosus*, "the Nilgiri race is a little larger and appears to
be a little darker in colour."   He gives as the measurements of a specimen, " beak ·4, wing 2·2, tail 1·75, tarsus ·7," and says "the black
lores appear more developed in the Nilgiri bird."   These observations
evidently refer to *Z. palpebrosus* ; but it appeared to me desirable to

mention them in my account of *Z. ceylonensis* for the purpose of showing the difference between the hill species of the two countries. I believe Dr. Jerdon is under the impression that he has seen *Z. ceylonensis* in India ; but he has no record of it.

Z. ceylonensis differs somewhat in habits from *Z. palpebrosus*. It frequents hedges and bush-jungle rather than trees, clinging Tit-like to the stems, and often covering its forehead with pollen from the flowers which it busily examines for insects. As these birds are very common and constantly flying in small parties from bush to bush, uttering their lively chirp, they attract attention ; and the little " White-eye" is familiar to most Europeans who visit Nuwara Eliya. In the winter the males associate in flocks of fifteen or twenty ; and it is then rare to find a female in their company. I believe the latter are for the time solitary, as, with one exception, the numerous specimens I have shot from different flocks have proved to be males. The breeding-season is probably about April or May ; but I have been unable to obtain any particulars of their nesting.

The distinction between the two species of *Zosterops* found in Ceylon will be readily seen on reference to Plate XX.

182. PARUS CINEREUS, Vieill.

Very abundant at Nuwara Eliya and on the upper hills at all seasons, and found occasionally on the western coast, around Colombo and not far from Galle. Layard says it is " not uncommon throughout the island," but I have never seen it in the Aripo district or in the extreme south. Like many hill birds it is often met with near Kandy ; but I expect its appearance about Colombo and in some other parts of the low country is exceptional, as when found there it is by no means numerous. It has the usual habits of the Titmouse family.

Bill black ; irides black ; feet leaden.

Ceylon, India (except Bengal), Malaya.

183. CORVUS LEVAILLANTI, Less.

Corvus culminatus, Sykes.

General in the low country, and especially frequenting native villages and the more uncultivated districts in the interior. It is rare at Colombo compared with *C. splendens*, and was not so numerous as that species at Aripo. I believe Crows are unknown on the upper hills ; but I have heard of their having been occasionally seen for a a day or two on coffee-estates 3000 or 4000 feet high.

Bill black ; irides dark brown ; feet black.

Ceylon, India to the Malay peninsula.

184. CORVUS SPLENDENS, Vieill.

This well-known bird is much more numerous on the coast generally than inland, and is found in great abundance in all the large towns, but is not met with in native villages so much as the last species. It was common at Aripo ; and at Colombo it is very abundant, not confining itself to the shore, but boarding the vessels as soon as they are anchored in the harbour or roadstead, paying frequent visits while they remain there, and only reluctantly leaving them at their departure when they are two or three miles away. It is unnecessary to say more of the well-known inquisitive, thievish habits of these birds than that in Ceylon they fully keep up the character they have obtained elsewhere. From the comparative localization of this bird in the larger towns in the south-west of Ceylon, Mr. Hugh Nevill has stated (Journ. Roy. Asiat. Soc., Ceylon Branch, 1870–71, p. 33) that "there is no doubt it is not indigenous to the south of the island, having been introduced by the Dutch at their various stations as a propagator of cinnamon, the seeds of which it rejects uninjured." I have been unable to discover on what evidence this statement has been made. This Crow has certainly been in Ceylon long enough to spread over every part of the island if its habits or inclinations had led it to do so ; but on both sides of the island it is comparatively local ; and whilst on a coasting voyage from Ceylon to Calcutta, and calling at numerous places on my way, I found on the Indian coast the same localization of this bird in the larger ports as is the case in Ceylon.

The Ceylon birds are smaller than those in India, and, according to Blyth, are darker, but I have not had an opportunity of comparing a sufficient number of specimens from the two countries to be able to judge on this point. Jerdon says nothing of the neck changing from ashy to a dull fawn-colour in old birds in India ; but this is the case in Ceylon. The young birds are very dark on the neck ; and these may possibly have been the subjects of Blyth's observations. Specimens of this Crow from Ceylon and India are now, however, in the Gardens of the Society, and will afford ready means of comparison of any changes that may take place.

Bill black ; irides brown ; feet black.

Ceylon, India, Assam, Burmah ?

185. CISSA ORNATA, Wagler.

Peculiar to Ceylon. This remarkably handsome species has attracted some attention since it was described by Blyth as *C. puella* from specimens forwarded by Layard ; but it had been previously made known by Wagler. It is, so far as is known, essentially a hill bird, found most abundantly at about 5000 feet and upwards, but at certain seasons descending as low as 1500 feet. This is about the elevation of Kandy ; and the jungles in the immediate neighbourhood of that city, nearly in the centre of the island, appear to be the lower limit of the range of this and many other hill species. In the cold season, which is only really perceptible on the hills, these birds are numerous at Nuwara Eliya, frequenting the dense bushes growing under the trees in forest-jungle. They are very noisy, continually uttering a harsh Jay-like scream, both when perched and flying. There is consequently little difficulty in finding them out when they are in the neighbourhood ; but from their keeping so much to the dense jungle I have on several occasions worked my way quietly through the bushes to within a few yards of the birds without being able to get sight of them. The specimens I obtained were apparently not quite mature, as the blue of the underparts was not uniformly developed ; otherwise they were in good feather, and enable me to give a description of the species.

Whole head and neck rich deep chestnut; back, tail, and underparts (in adults) bright cobalt-blue, the tail-feathers tipped and more or less margined externally with white ; quills light chestnut on the outer webs, black on the inner.

Bill red (adult), tipped black (young) ; irides light brown ; feet coral-red.

Ceylon hills.

186. ACRIDOTHERES TRISTIS, Linn.

Very common in the low country, and generally distributed. They were very numerous at Aripo ; and a young bird brought to me by some natives soon became tame enough to be allowed its liberty in the house, sometimes escaping through the window to the adjoining trees, but always allowing itself to be caught, or going into its cage when held up to it. It became rather troublesome at last from its fondness for standing on the top of my head or perching on my hand when I was writing or engaged in some other work at the table.

Ceylon specimens are much darker than those obtained in India.

Bill yellow ; orbits yellow ; irides dark brown ; feet pale yellow.

Ceylon, India, Assam, Burmah.

187. TEMENUCHUS PAGODARUM, Gmel.

Obtained by Layard in the north of the island, and by Kelaart at Trincomalie.

Ceylon, India.

188. TEMENUCHUS SENEX, Temm.

Peculiar to Ceylon ; described by Layard as *T. albofrontatus*, as it was believed to be new; it has since been recognized as *T. senex*, Temm., erroneously described by Bonaparte as from Bengal   Several specimens have been received by Lord Walden, of which, however, only one has the head entirely grey, the true character of *T. senex*. Layard gives the following description of his bird, which is now in the British Museum :—

" General colour of back, tail, and wings black with a green gloss ; forehead albescent ; hinder feathers of crest brownish black with albescent shafts ; general colour of breast, throat, vent, and under tail-coverts albescent, the shafts of the feathers on the throat shining white."

It is, I believe, from the lower hills, and appears to be rather a local species.

Ceylon.

189. PASTOR ROSEUS, Linn.

Layard " found large flocks of these birds " quite at the north of the island in July, but did not see them afterwards. They have also been obtained at Putlam, on the west coast; and I have little doubt that I saw a flock at Aripo in 1866, but I could not get near them. It is rather remarkable that this bird is not better known in Ceylon, as in India, according to Jerdon, it is most abundant in the south and south-west.

Ceylon and India westward.

190. EULABES RELIGIOSA, Linn.

Recorded by Layard as common on the west coast. I have never met with it at Aripo, and believe it is more frequently seen in the south. There are many Ceylon specimens in Lord Walden's collection, most probably procured in the south-east of the island.

Ceylon, South India.

191. EULABES PTILOGENYS, Blyth.

This well-marked species of hill Myna is peculiar to Ceylon, and is found in flocks on the upper hills chiefly, but sometimes met with in the neighbourhood of Kandy. It frequents the tops of the trees; and at Nuwara Eliya, where it is often numerous, I have found it wild and difficult of approach. I have heard, however, of large numbers having been killed on some of the coffee-estates in the early morning or evening. Its call is constantly repeated when on the wing, and sometimes when perched on the tops of the trees.

This species may be readily distinguished by the yellow lappets at the back of the head, and the absence of any naked skin about the eye and cheeks.

Bill deep orange, base black; irides brown, lappets yellow; feet dull yellow.

Ceylon.

192. PLOCEUS BAYA, Blyth.

This bird, called by Layard P. *philippinus,* is said by him to be migratory and to breed in June. It was, however, generally to be found at Aripo; and there they used to build their curious nests in December on the trees close to my house. A young bird was brought to me in February which was just ready to leave the nest. I have never seen the nest of this species in any other than ordinary branching trees; but Layard says it builds on palms and other trees indiscriminately.

Ceylon, India, Assam, Burmah, Malaya.

193. PLOCEUS STRIATUS, Blyth.

I have not met with this species at Aripo or on the west side of Ceylon; and Layard, who found it on the east side, thinks it is confined to that part of the island. It is rather remarkable, however, that this bird should not change its quarters according to the season and, like many other species, migrate from one side to the other at the change of the monsoons.

This is the species most probably given by Layard under P. *manyar,* Horsf., which is a Javan bird.

Ceylon, North and Central India, Burmah, parts of Malaya.

194. MUNIA MALACCA, Linn.

195. MUNIA RUBRONIGRA, Hodgs.

196. MUNIA UNDULATA, Latham.

197. MUNIA STRIATA, Linn.

198. MUNIA MALABARICA, Linn.

With the exception of *M. rubronigra*, which I have not seen, and was recorded by Layard only from Galle, the above species are more or less abundant in the low country—*M. undulata* and *M. malabarica* being the most numerous, and the former perhaps the most widely distributed. I have seen many nests of *M. undulata* at Aripo and near Colombo, and have often watched the birds biting off the grass-stems and taking them to the nest, which has been generally a large structure, sometimes placed near the end of a branch, but more commonly in a thick bush.

These species are more or less distributed through India and the neighbouring countries eastward of it.

199. MUNIA KELAARTI, Blyth.

Peculiar to Ceylon, and confined to the upper hills. It is abundant at Nuwara Eliya at all seasons, frequenting the gardens and cultivated ground, and may often be seen on the roads feeding, like the Sparrows, on what it can find there. I have specimens in all stages of plumage. The adult bird may be distinguished from *M. pectoralis*, Jerdon, with which it was at first confused, by its having the rump and underparts, from the breast downwards, brownish black, with each feather centred, barred, and margined with white, producing a mottled effect ; the under tail-coverts are only centred white ; and the extremity of the upper tail-coverts is tinged with glistening yellow. Young birds have the throat speckled brown and white, and the underparts faintly mottled with two shades of light yellowish brown.

Bill lead-colour, very dark in adults ; irides brown ; feet leaden.
Ceylon.

200. ESTRELDA AMANDAVA, Linn.

I have seen specimens of this bird which were procured by Mr. Legge from a grass-field adjoining his house at Colombo. It had not been previously observed in Ceylon ; and it may be, as Mr. Legge suggests as possible (J. R. A. S., C. B., 1870-71, p. 53, note), that some of the many birds of this species imported into Ceylon have escaped from confinement and become acclimatized. The occurrence of *Munia rubronigra* (a North-Indian species) only about Galle may perhaps be accounted for in the same manner, if no mistake was made in its identification.

Ceylon, India, Assam, Burmah.

201. PASSER INDICUS, Jard. & Selby.

Found in Ceylon wherever there are human habitations. It is

abundant at Nuwara Eliya, but I was told by old residents that
they remembered the time when the now common Sparrows and
Musquitos were unknown at that elevation.
Ceylon, India, eastward to Siam.

202. MIRAFRA AFFINIS, Jerdon.

I have found this species common at Aripo ; and Layard has recorded it from the north also.   I am not sure that it does not also
occur at Colombo.
Bill dusky above, pale brown below ; irides brown ; feet fleshy
brown.
Ceylon, South India, Upper Burmah.

203. PYRRHULAUDA GRISEA, Scop.

Confined to the northern part of the island.   Layard believed it
was migratory ; but I have seen it at Aripo at all seasons, in pairs
during the summer, and in flocks during the winter months.
Bill pale brown ; irides brown ; feet fleshy.
Ceylon, India westward to Arabia.

204. ALAUDA GULGULA, Frankl.

Very common in the low country ; but I have no recollection of
seeing it on the hills.   It has, however, been recorded, I believe, from
the upper country by Kelaart.   It was abundant at Aripo.
Bill dusky above, paler below ; irides brown ; feet fleshy brown.
Ceylon, India.

205. CROCOPUS CHLOROGASTER, Blyth.

I have obtained this Pigeon near Aripo ; and it is said by Layard
to be confined to the north of the island.
Ceylon, South and Central India.

206. OSMOTRERON BICINCTA, Jerdon.

This species is also found at times in wild jungle south of Aripo.
I have likewise met with it a few miles from Coombo ; but it is
recorded as more numerous further south.
Ceylon, India eastward to Tenasserim.

207. OSMOTRERON POMPADOURA, Gmel.

The description of this species given by Gmelin was from a drawing of a Ceylon bird.   Layard believed it to be a variety of O. malabarica, Jerdon ; and Blyth has since given it the name of flavo-
gularis ; but the difference between Blyth's species and O. pompadoura can only be traced in the under tail-coverts, and there is a
variation in this difference.   O. pompadoura and O. flavo-gularis
agree precisely in differing from O. malabarica in having the head
less grey and the throat more yellow, and in not having the under
tail-coverts cinnamon ; this colour, however, Mr. Blyth tells me is
only found in the male of O. malabarica.   Specimens of O. pompa-

PROC. ZOOL. SOC.—1872, No. XXX.

doura and *flavo-gularis* in Lord Walden's collection are undistinguishable, except in having the under tail-coverts green margined with white, entirely white, or white margined with yellow. Lord Walden's opinion that these differences are due to season or age appears to me likely to be correct; if not, the number of species founded on the colour of these particular feathers will have to be increased.
Ceylon, South India.

208. CARPOPHAGA SYLVATICA, Tickell.

Recorded by Layard as mostly found on the mountain-zone. He mentions it under the name of *C. pusilla*, Blyth; but the difference in size of this Pigeon from Ceylon and parts of India is not generally recognized as of specific value.
Ceylon, India to Burmah, and Hainan.

209. ALSOCOMUS PUNICEUS, Tickell.

This Pigeon, known to the Singhalese by a name literally translated "Season Pigeon," is recorded by Layard only as a rare visitor; and, according to the natives, "it appears during the fruiting of the cinnamon-trees." I have never seen it.
Ceylon, Eastern side of Central India, Assam, Arrakan, and Tenasserim.

210. PALUMBUS TORRINGTONIÆ, Kelaart.

Peculiar to Ceylon. It is found in great abundance on the hills, but changes its locality according to the season and the time at which the fruit of particular trees ripens. I have found it numerous at Nuwara Eliya at the end and beginning of the year; and it is occasionally found there at other times. It is allied to *P. elphinstonei*, Sykes, but differs essentially from it in having the back and wings dark slaty, and the underparts strongly vinaceous. It is known on the hills as the "Blue Pigeon."
Bill dusky, tip pale green; irides dark yellow; feet fleshy red.
Ceylon.

Macropygia macroura (Gmelin). With reference to the occurrence of this species in Ceylon, as stated by Bonaparte, Lord Walden has been good enough to send me the following note, with permission to make use of it:—

"The titles *Columba macroura*, Gmel. (1788), and *Columba macerona*, L. S. Müller (1776), were founded on the *Tourocco* of Buffon (Hist. Nat. Ois. ii. p. 553, and Pl. Enl. 329). Buffon figured this Pigeon from a Senegal example, presented by Adanson under the name of *Tourterelle à large queue du Sénégal*. But he afterwards (Hist. Nat.) substituted for Adanson's title that of *Tourocco*, because, as he says, while Adanson's bird possessed many of the characters of the European Turtledove, it carried its tail like 'le Hocco' (*Crax*). *Tourocco* may therefore be translated Turtledove-Curassow. Buffon is most circumstantial in his account of the locality whence his bird was obtained; and the fact that the specimen bore a title given

by Adanson strongly corroborates the Senegal origin.   Yet Bonaparte (Consp. ii. p. 57) says 'ex Ceylon, nec Senegal.'   The Prince
was also (*l. c.*), I believe, the first who referred *C. macroura*, Gm.,
to the genus *Macropygia*.   Still it is doubtful whether he ever
saw an example of the bird, and the diagnosis given by him of the
species only contains the prominent characters discernible in the
plate quoted."

211. COLUMBA INTERMEDIA, Strickl.

There are two stations on the Ceylon coast which "Rock-Pigeons"
are known to frequent.   The principal one is Pigeon Island, a large
mass of isolated rocks well known on the east coast, and about
eighteen miles north of Trincomalie.   I have visited this locality;
and I have no doubt that Pigeons, probably of this species, are
found there at a particular season of the year, according to the
general report of the natives on the adjoining mainland; but I did
not see any when I was there.   Layard mentions their having been
killed about fifty miles inland from Trincomalie.   The other station
is off Berberyn, not far from Galle.
Ceylon, India to Burmah.

212. TURTUR RUPICOLA, Pall.

Layard records having shot a young bird of *T. orientalis*, Lath.
This may be the above species; but I am disposed to think his identification doubtful, as his only specimen was a young bird.

213. TURTUR SURATENSIS, Gmel.

Very common in the low country, and abundant at Aripo.
Ceylon, India.

214. TURTUR RISORIA, Linn.

Very numerous in the north, and, I believe, not uncommon
throughout the low country.
Ceylon, India.

215. CHALCOPHAPS INDICA, Linn.

This handsome Dove is found in all parts of the island except the
north.   I have met with it in cultivated districts near Colombo and
in the extreme south, on the road through the forest between Kandy
and Trincomalie, and at Nuwara Eliya, where at the end of the year
it frequents the jungle in great numbers.   It has a low rapid flight,
and a peculiar moaning coo, more like the note of some Owls than
that of a Dove.
Ceylon, India, eastward to Tenasserim.

216. PAVO CRISTATUS, Linn.

Common in all jungly districts within a moderate distance of the
coast.   So far as my observations and inquiries have gone, it is unknown in the hill-country; and it is more numerous in the eastern

and northern parts of the island than in the more cultivated south and west.
Ceylon, India.

217. GALLUS STANLEYI, Gray.

The Ceylon Jungle-fowl is remarkable not only for being peculiar to the island, but also for being common in all parts of it where the country is uncultivated and there is jungle of a moderate height. Although especially abundant in the low country, it is often very numerous even on the upper hills, and is attracted to the particular localities where the "*nilloo*," the native name for some species of *Strobilanthes* growing at 5000 feet and upwards, is at the time in seed. I have entirely failed to discover that any thing is known among botanists of the seeds of the *Acanthaceæ* possessing narcotic or other poisonous properties; but it is well known that the Jungle-fowl after feeding for a time among the *nilloo* become partially blind or stupified, so that they may frequently be knocked down with a stick. This stupefaction is generally attributed to the *nilloo*-seeds, which are so largely eaten by these birds; but in the absence of any known poisonous properties in these seeds, it appears possible that the birds may really suffer from devouring some fungus or other plant found in the damp woods where the *nilloo* grows.

At daybreak the crow of the Jungle-cock is first heard; and for an hour or two after sunrise, if the birds are at all numerous, they may be heard challenging each other on all sides. On these occasions a successful shot may sometimes be obtained by remaining perfectly still between two birds which are challenging and gradually approaching each other. Some of the native hunters are very expert in calling the Jungle-cocks, by beating on a loose fold of their cloth, so as to produce an imitation of the sound of a bird's wings just as it is alighting: no time must be lost with the gun on these occasions, as the cocks discover the deception the moment they get sight of you, and instantly run off with drooping tails like Pheasants. It is not difficult in favourable jungle to approach a calling bird within easy shot; and under these circumstances I have generally found the cock strutting up and down a low horizontal branch of a tree, raising and lowering its head, and every now and then giving utterance to its peculiar crow, which has been likened to the sound of "George Joyce." When the bird is tolerably close, the syllable "ck" is heard preceding those two sounds, which are so familiar to persons who have been wandering in the jungles of Ceylon. In some of the wilder jungle-roads, a cock and hen may sometimes be seen feeding together; but generally the hens are very shy, and not many of them are killed.

Mr. Layard tells me that there is no doubt about this Junglefowl sometimes breeding with the domestic poultry in the native villages. I have seen young Jungle-fowl, which had been hatched under domestic hens, running about with the other chickens; but they were always rather wild and invariably roosted out of doors; and those which were not sooner or later killed by some accident, ultimately took to the jungle. Some others, which, however, were reared in an aviary at Colombo by my friend Dr. Boake, became quite tame, and were in good feather when he kindly allowed me to send them to London for the Society's Gardens ; but they all died when they were almost within sight of England.

Mr. Blyth can hardly be correct in his description of the head and appendages in this species. He says (Ibis, 1867, p. 307), " The cock has a yellow comb with a red edge, and the cheeks and wattles (as I remember them in the living bird) are chiefly yellow." His description of the colour of the comb is approximately correct, as the extent of the yellow varies in different specimens; but I am too familiar with the appearance of the living or freshly killed bird to have any doubt about the cheeks and wattles being red. These parts assume a dark livid appearance a few hours after death ; but the yellow in the comb remains, and is evident even in old dry skins. The size of the comb and wattles varies, and probably depends on age.

The following details were taken from a fine adult cock I killed at Aripo, and were noted down on the spot :—

Bill brown, front of the lower mandible pale yellow ; irides buff ; comb, wattles, and naked skin about the head purplish red, the comb having a large wing-shaped spot of yellow occupying the middle of the posterior half, very bright at its origin immediately over the eye, and shading off at its margin into the colour of the comb ; feet and legs pale yellow.

Ceylon.

218. Galloperdix bicalcarata, Forst.

Peculiar to Ceylon ; abundant on many parts of the hills, and frequenting also jungly places in the low parts of the southern half of the island. During the winter months it is numerous in the coffee-districts and upper hills, and is trapped in large numbers by the natives. It is skulking in its habits and difficult to flush, usually seeking concealment in the thicker parts of the jungle when it is disturbed. They bear confinement well in Ceylon ; but some specimens I brought to England, although apparently strong and well on their arrival, all died within three days after the ship entered the Thames.

Bill red ♂, dusky ♀; irides brown ; feet fleshy red.

Ceylon.

219. Francolinus pictus, Jard. & Selby.

The occurrence of this species, said to have been well identified, was noticed three or four years ago in one of the Colombo newspapers. I did not see the specimens, and I cannot now give the precise date or particulars of where they were obtained.

Ceylon, Central India.

220. Ortygornis ponticeriana, Gmel.

Common in the north of Ceylon, and found also in the cinnamon-gardens at Colombo. These birds may have escaped from confinement, as large numbers of them are brought alive to the Colombo market from Tuticorin on the Indian coast; Mr. Legge, however, has also seen the bird at Galle. This species is indigenous in the north, and is always very abundant at Aripo. The large compound surrounding my house at that place was virtually nothing but a considerable piece of jungle fenced in, and was frequented by many kinds of wild animals and birds. Partridges were very numerous there; and they might be seen or heard at all hours of the day, and often within a few yards of the house. They roosted in low bushes.

Bill dusky; irides brown; feet dull red.

Ceylon, South, Central, and North-west India, Persia?

221. PERDICULA ASIATICA, Lath.

Layard mentions having seen a pair of these birds which were caught alive near Colombo. He speaks of it under the name of *P. argoondah*, Sykes.

Ceylon, South India.

222. EXCALFACTORIA CHINENSIS, Linn.

I have seen this bird from Kandy and the cinnamon-gardens at Colombo; and Layard says it is common in the south.

Ceylon, India, eastward to China, Malaya, Australia.

223. TURNIX TAIGOOR, Sykes.

Common in all parts of the low country. I have found its eggs at Aripo in February.

Bill lead-colour; irides pale yellow; feet pale leaden.

Ceylon, India.

224. CURSORIUS COROMANDELICUS, Gmel.

I believe the Indian Courser is resident in the north of Ceylon, as I have seen it in almost every month of the year at Aripo. It is more numerous, however, in the winter months, being then in small parties of six or eight. Its flight is heavy and flapping, like that of the Lapwings; but it runs lightly and fast; and when separated from its companions, I have more than once seen it running along behind the bund of a dry paddy-field, with head lowered and wings trailing on the ground, presenting a most curious appearance, as the colour of the back resembled that of the dry mud, and there was nothing to attract attention but the drooping black primaries. Layard appears to have occasionally met with this bird, but only in April.

Bill black; irides dark brown; feet cream-colour.

Ceylon; Central and West India.

225. CHARADRIUS FULVUS, Gmel.

Charadrius longipes, Temm., apud Jerdon.

The Ceylon birds have the ash-coloured axillary plume characteristic of this species; they are migratory, appearing at Aripo in August, many of them then having some remains of the black

breeding-plumage. Throughout the winter they are abundant in
the north, and are occasionally seen as far south as Colombo, frequenting the esplanade with some of the smaller Plovers.

Bill black ; irides brown ; feet dark leaden.

Ceylon, India, Eastern Asia to Australia, and Polynesia.

226. ÆGIALITES MONGOLICUS, Pall.

Mr. Edmund Harting, in a series of exhaustive papers "On
rare or little known Limicolæ" ('Ibis' 1870), has worked out the
synonymy of this species, among others, and identified the bird
given by Jerdon under *Æ. pyrrhothorax*, Temm., with that described by Pallas. It is doubtless the one mentioned by Layard as
Hiaticula leschenaultii, Less., as I have no reason to think the
much larger *Æ. geoffroyii*, Wagler, is found in Ceylon.

Æ. mongolicus is a winter visitor to Ceylon, and is then very
abundant on the coast, commonly associating with *Æ. cantianus*.
All the specimens I have examined have been in winter dress.

Bill black ; irides dark brown ; legs grey, feet dark grey.

Asia to North Australia.

227. ÆGIALITES CANTIANUS, Lath.

Mr. Harting has been good enough to examine my specimens of
this and the preceding species, and he tells me that a small Plover
which I had been unable to identify is the young of *Æ. cantianus*.
I obtained specimens of this species in different states of plumage ;
but the greater number of these birds found in Ceylon are young ones,
and apparently diminutives of *Æ. mongolicus*. I have occasionally
got specimens in nearly, if not quite, full plumage.

Bill black ; irides dark brown ; feet dark grey, legs paler (in the
young).

Europe, Asia.

228. ÆGIALITES DUBIUS, Scop.

Ægialites philippensis, Scop., apud Jerdon.

This well-known little Plover is common in Ceylon, and, I believe,
resident there, as it is certainly found during a great part of the
year at Aripo.   Although associating to some extent with the other
Sand-Plovers, it does not always keep with the party, but wanders
off to some distance when feeding. It is particularly fond of standing on any little natural elevation of the ground or heap of rubbish
on the beach.

Bill black ; irides dark brown ; feet yellow.

Ceylon, India, eastward to China and Japan.

229. CHETTUSIA GREGARIA, Pall.

I have identified a single specimen of this Plover shot by Mr.
Bligh on the Galle face at Colombo. It has not been before observed
in Ceylon.

Ceylon, parts of India, West Asia, and South-east Europe.

230. LOBIVANELLUS INDICUS, Bodd.

Lobivanellus goensis, Gmel., apud Jerdon.

Found in all open parts of the low country, and generally in pairs. Resident in Ceylon.

Ceylon, India.

231. SARCIOPHORUS MALABARICUS, Bodd.

Sarciophorus bilobus, Gmel., apud Jerdon.

Distribution much the same as that of the last species, but it is more numerous in the north. It was always abundant at Aripo, and was found in large flocks during winter. Jerdon, in his description of this species, has omitted to mention that the chin and upper part of the throat are dull black. This appears as soon as the young are well able to fly, and remains at all seasons. It is present in all the Indian specimens I have seen.

Bill yellow, tip black; irides pale yellow; wattles yellow; feet yellow.

Ceylon, India.

232. ESACUS RECURVIROSTRIS, Cuv.

I have only seen this bird occasionally in the Aripo district. It was usually in pairs on the banks of the Aripo river. I have shot this bird in August, from which it would appear to be a resident.

Bill greenish yellow, tip black; irides pale yellow; feet yellow.

Ceylon, India.

233. ŒDICNEMUS CREPITANS, Temm.

Common in the north at all seasons. I have also flushed it in the cinnamon-gardens at Colombo.

Asia, N. Africa, Europe.

234. STREPSILAS INTERPRES, Linn.

I obtained one specimen in August on the coast a few miles north of Aripo. Layard also met with it in the north, and once at Colombo. It is rather a scarce bird in Ceylon.

World-wide distribution.

235. DROMAS ARDEOLA, Paykul.

I have never seen this remarkable bird; but specimens were obtained by Layard—at sea, with one exception. He follows Blyth in placing it near the Terns.

Ceylon, India, Red Sea.

236. HÆMATOPUS OSTRALEGUS, Linn.

Layard records having seen one or two of these birds near Jaffna.

Ceylon, Indian and European coasts.

237. SCOLOPAX RUSTICULA, Linn.

The occasional appearance of the Woodcock on the Ceylon hills has been reported on "sportsman's authority;" and it is now confirmed by Mr. S. Bligh, who writes to me that he has just examined a specimen quite recently killed at Nuwara Eliya.

238. GALLINAGO NEMORICOLA, Hodgson.

239. GALLINAGO STENURA, Temm.

240. GALLINAGO SCOLOPACINA, Bonap.

241. GALLINAGO GALLINULA, Linn.

Of these four reputed Ceylon species *G. stenura* appears to be the only one which has been positively identified. It is the Common Snipe of sportsmen; and I believe there are but few persons in the island who are aware of the peculiarity in the tail-feathers by which it can be at once distinguished from *G. scolopacina,* which it is generally believed to be. *G. stenura* is found all over the island in the winter months; and although of course much more abundant in paddy-growing districts, it is also numerous in swampy plains on the upper hills. *G. nemoricola* was recorded from Nuwara Eliya by Mr. Hugh Nevill as new to the island (J. R. A. S., C. B., 1867–70, p. 138); but although he, I believe, examined the specimen, the skin was not preserved, and he himself told me that he identified the bird, after he had left the hills, by the coloured figure in Jerdon's 'Illust. Ind. Ornith.,' from which work he has evidently taken his description of the species. Although neither Layard nor Kelaart mentions this bird, Jerdon speaks of it as being found on the "elevated regions of Southern India and Ceylon," but does not give any authority. In the case of *G. gallinula,* Layard thought that "sportsman's authority" might be trusted, as the "Jack" would not be easily confounded with the other Indian Snipes, and he had been informed by a person likely to be acquainted with it that it was not uncommon in the north a few years previously. None of these species is unlikely to occur in Ceylon; but, except in the case of *G. stenura,* the evidence in their favour is not quite as clear as could be wished.

242. RHYNCHÆA BENGALENSIS, Linn.

Not uncommon in the low country during the winter. Layard says some remain to breed, "the season of incubation being from May to July." He tells me that he obtained many eggs of this species. Jerdon also gives June and July as the breeding-time of this bird in India. It apparently varies, however, as a bird caught near Colombo, and sent alive to me on the 31st of December, was found to have laid an egg in the basket in which it was packed. This egg has been identified by Mr. Layard as that of the Painted Snipe, although its ground-colour is rather paler than usual.

Ceylon, India, Burmah to S. China, Africa.

243. LIMOSA ÆGOCEPHALA, Linn.

Recorded by Layard. I have not met with it.

Ceylon, parts of continental Asia and Europe.

244. TEREKIA CINEREA, Gmel.

I obtained one specimen in winter plumage, out of a flock of five, in April 1869; they were in a small swamp near the sea at Aripo. It appears to be new to Ceylon.

Bill dusky, base yellow; irides brown; feet pale orange.
Europe, Asia to Australia.

245. NUMENIUS ARQUATA, Linn.?

The Ceylon Curlew requires further examination; it may prove to be *N. lineatus*, Cuv., which Mr. Blyth tells me is commonly found in India.

246. NUMENIUS PHÆOPUS, Linn.

This and the preceding species are common on many parts of the coast, and were often found at Aripo, but never in flocks.
Europe, Africa, Asia.

247. TRINGA SUBARQUATA, Gmel.

I have obtained this bird in May at Aripo with the breeding-plumage far advanced.

248. TRINGA MINUTA, Leisler.

This was the common Stint on the shore at Aripo, yet it appears doubtful whether it is found in India (Jerdon, Birds of India, App. p. 875).

Bill black; irides brown; feet leaden-black.

249. TRINGA SALINA, Pall.

Tringa subminuta, Midd.

I obtained two of this species at Aripo, in January 1870, for the first time. It is new to Ceylon, although Blyth, as quoted by Jerdon (*ut suprà*), states that it is the common Little Stint of India. My specimens of these two Stints have been carefully examined and identified for me in England.

Bill black; irides brown; feet dull olive.

250. TRINGA PLATYRHYNCHA, Temm.

This appears to be rare in Ceylon. Layard only obtained one two specimens quite in the north.

251. ACTITIS GLAREOLA, Gmel.

Exceedingly abundant in all wet places. I have counted twenty round a small pool in my compound at Aripo during the rains.

252. ACTITIS OCHROPUS, Linn.

Numerous, but less so than the last species.

253. ACTITIS HYPOLEUCOS, Linn.

Very common in all parts of the low country, and less so on the hills. I have seen it as high as Nuwara Eliya in February. It is probably resident in Ceylon.

254. TOTANUS GLOTTIS, Linn.

255. TOTANUS STAGNATILIS, Bechst.

Both very common at Aripo, and generally so in the low country.

256. TOTANUS FUSCUS, Linn.

257. TOTANUS CALIDRIS, Linn.

These species were considered common by Layard ; but I have not seen them.

258. HIMANTOPUS AUTUMNALIS, Hass.

Himantopus candidus, Bonn.

Not uncommon at Aripo during the rains.

259. RECURVIROSTRA AVOCETTA, Linn.

Two of these birds, killed near Jaffna, are recorded by Layard.

Almost all these small Waders are, I believe, winter visitors to Ceylon after breeding in Northern Europe or Asia.

260. HYDROPHASIANUS CHIRURGUS, Scop.

Very common in the neighbourhood of Colombo. Beautiful specimens in various states of plumage are sometimes brought in for sale by the Singhalese, who walk through the flooded marshes and wait patiently, with the water often above their waists, till they can make sure of a successful shot. I have not seen this bird in the north ; but, as Layard mentions, it may be sometimes observed walking on the lotus-leaves in the lake at Colombo.

Bill bluish, tip green ; irides red-brown ; feet leaden.

Ceylon, India, China.

261. PORPHYRIO POLIOCEPHALUS, Lath.

Common in suitable situations, but shy and fond of concealment. They are numerous in the neighbourhood of Colombo.

Ceylon, India.

262. GALLICREX CRISTATUS, Lath.

This bird is also common about Colombo and in marshes in the south.

Ceylon, parts of India, Burmah, Malaya, China.

263. GALLINULA CHLOROPUS, Linn.

Layard met with one specimen of this bird in the north ; but I have not heard of any others, although it appears to be general in India.

Europe, Asia, Africa.

264. GALLINULA PHŒNICURA, Forster.

This well-known species was first described and characteristically figured by Forster (1781) from a Ceylon specimen. It is very common in suitable situations throughout the low country; but I am not aware that it is found on the hills, except near their foot.

In what appears to be rather an old bird the upper part of the back is irregularly barred with grey, and the chestnut is confined to the sides of the rump and under tail-coverts. This was a male, and when shot was in company with a presumed female and a small black chick. Jerdon does not mention the characteristic white face and forehead in his description of this species; and I observe in the British Museum a specimen from Malaya labelled *G. phœnicura* in which the white is confined to the underparts. It appears to have rather a stouter bill, and may be a distinct race or species, possibly the one from which Jerdon took his description, in which he says "irides blood-red, legs green." These chraacters do not agree with the following in true *G. phœnicura* from Ceylon :—

Bill green, ridge dull red ; irides brown ; legs and feet light yellow-brown.

Ceylon, India to Malaya, S. China, Formosa.

265. PORZANA PYGMÆA, Naum.

Layard records having obtained one specimen.
Ceylon, India, China, Japan.

266. PORZANA FUSCA, Linn.

Recorded by Layard as rare.
Ceylon, India, E. Asia.

267. RALLINA CEYLONICA, Gmel.

This bird arrives in Ceylon in October, just at the change of the monsoon, and takes refuge in the first place of concealment it can find, often entering the houses and hiding amongst the furniture. I have caught the bird under these circumstances at the hotel at Colombo. Although this Rail is only a winter visitor to Ceylon, specimens of it from India appear to be rare, and the North-Indian race has been separated by Blyth under the name of *R. amauroptera*. The distribution of the true *R. ceylonica* appears to be uncertain.

Bill dusky above, green below ; irides red-brown ("carmine, with an inner circle of yellow," *Layard*) ; feet leaden brown.
Ceylon, S. India.

268. RALLUS STRIATUS, Linn.

Ceylon, India, Burmah to Malaya, Formosa.

269. RALLUS INDICUS, Blyth.

Ceylon, India, Tientsin.
These two species have both been recorded by Layard, but are said to be rare. *R. indicus* is very close to *R. aquaticus* of Europe, but

has been separated by Blyth; I have not had an opportunity of comparing them.

270. LEPTOPTILOS JAVANICA, Horsf.

I have seen this Stork close to Aripo and a few miles from Trincomalie, on both occasions in small parties. I believe it is a winter visitor and that it is only found in the northern half of the island, although by no means uncommon in particular districts.

Ceylon, India, Burmah to part of Malaya, Hainan.

271. MYCTERIA AUSTRALIS, Shaw.

Layard mentions having seen this bird near Jaffna; but I have never met with it.

Ceylon, India to Australia.

272. CICONIA EPISCOPUS, Bodd.

Ciconia leucocephala, Gmel.

Described by Layard as common in swampy lands; and although I have not met with the bird, it appears to be well known in suitable situations.

Ceylon, India to Malaya.

273. ARDEA CINEREA, Linn.

This Heron, considered by Layard to be very rare, is not at all uncommon in Ceylon. I have seen it in many parts of the island, and have had an opportunity of examining young birds on more than one occasion.

Asia, Africa, Europe.

274. ARDEA PURPUREA, Linn.

More common than the last species; it is very numerous in the south, and breeds near the Amblangodde Lake, a few miles from Galle.

Asia, Africa, Europe.

275. HERODIAS ALBA, Linn.

276. HERODIAS EGRETTOIDES, Temm.

Ardea intermedia, Wagler, apud Layard.

277. HERODIAS GARZETTA, Linn.

278. DEMIEGRETTA ASHA, Sykes.

279. BUPHUS COROMANDUS, Bodd.

These five species are all said by Layard to be common, and to breed in Ceylon. I have no doubt he is quite correct. Egrets of different kinds are abundant in the swamps throughout the island; but as I brought home no specimens with me, I cannot be sure of the correctness of my identifications. I have occasionally seen specimens of one species at Nuwara Eliya, in October, with a black bill and greenish feet, probably *H. garzetta*; but these birds are mostly found in the low country. They appear to be all widely distributed.

280. ARDEOLA GRAYII, Sykes.

Ardeola leucoptera, Bodd., apud Jerdon.

Exceedingly common in Ceylon.

281. BUTORIDES JAVANICA, Horsf.

Common, and resident in the north. I have occasionally seen it near Colombo.

Bill black above, yellow below; irides yellow; feet yellow-green. Ceylon, India, Burmah to Malaya, China.

282. ARDETTA FLAVICOLLIS, Lath.

283. ARDETTA CINNAMOMEA, Gmel.

284. ARDETTA SINENSIS, Gmel.

Of these three species *A. cinnamomea* is the commonest. They appear to be confined to the southern half of the island, and are all found in the neighbourhood of Colombo. They range from India more or less eastward.

285. NYCTICORAX GRISEUS, Linn.

Not uncommon in suitable places.
Asia, Africa, Europe.

286. GOISACHIUS MELANOLOPHUS, Raffles.

I was fortunate in getting a specimen of this purely eastern Bittern at Aripo, in November 1866; it was hiding among some low bushes a few yards from my house, and was a female in immature plumage. My shot disabled but did not kill it, and it struck at me furiously with its bill as I endeavoured to extricate it from among the thorns, the neck-feathers being erected in true Bittern fashion. Layard first observed this species in Ceylon, and obtained two or three specimens near Colombo. It is remarkable that this common Malay species should not yet have been observed in India, as the birds obtained by Layard and myself (no others are recorded from Ceylon) were all found on the west coast, and my specimen was from that part of the coast where migrants from India generally appear first.

Ceylon, Malaya, Japan, Philippines.

287. TANTALUS LEUCOCEPHALUS, Forster.

Ceylon, India, Burmah, Amoy.

288. PLATALEA LEUCORODIA, Linn.

Asia, Africa, Europe.

289. ANASTOMUS OSCITANS, Bodd.
Ceylon, India.

290. THRESKIORNIS MELANOCEPHALUS, Linn.
Ceylon, India, Burmah, Arrakan, China?

These four species are resident in Ceylon, and abundant in many
localities, but, I believe, more numerous in the south and south-east
than elsewhere, except, perhaps, the last one, which Layard says is
common in the north and north-west. I have only seen it on two
occasions near Aripo. *T. leucocephalus* was first described from a
Ceylon specimen.

291. FALCINELLUS IGNEUS, S. G. Gmel.
Not uncommon near Aripo, and apparently confined to the north.
World-wide distribution.

292. PHŒNICOPTERUS ANTIQUORUM, Temm.
Phœnicopterus roseus, Pall., apud Jerdon.
I have seen this species occasionally at Aripo in October and November; they were in flocks of from twenty to thirty, and presented a remarkable appearance as they flew in single file, with
outstretched head and legs. Layard speaks of having seen them in
very large numbers on the north and east coasts.
Parts of Asia, Africa, and Europe.

293. SARKIDIORNIS MELANONOTUS, Forster.
Ceylon, India, Burmah.

294. ANSERELLA COROMANDELIANA, Gmel.
Ceylon, India, Burmah, Malaya.

295. DENDROCYGNA JAVANICA, Horsf.
Dendrocygna arcuata, Cuv.
Ceylon, India, Burmah, Malaya.

296. SPATULA CLYPEATA, Linn.
Asia, Europe.

297. ANAS PŒCILORHYNCHA, Penn.
Ceylon, India, Burmah.

298. DAFILA ACUTA, Linn.
Asia, Europe.

299. QUERQUEDULA CRECCA, Linn.
Asia, Europe.

300. QUERQUEDULA CIRCIA, Linn.
Asia, Africa, Europe.

These eight species are all recorded by Layard; and *S. melano-notus* was first obtained in Ceylon. The only one commonly distributed throughout the island is *D. javanica*; it breeds in many localities, and is known among Europeans as the "Teal." *A. coromandeliana* is tolerably numerous in the north and east, and, I am told, breeds near Battacaloa; it is sometimes found near Colombo. The others are mostly found in the north, where, according to Layard, *Q. crecca* and *circia* are very abundant in winter. Layard also mentions having seen on several occasions, through a telescope, what he believed to be *Branta rufina*, Pall.; but that species has not been yet identified from Ceylon.

301. PODICEPS PHILIPPENSIS, Bonn.

Very common on all large pieces of water, and often associated in flocks. I have counted thirty-eight together on the Colombo lake.

Bill black, tip white, base of lower mandible dull green; irides dark yellow; legs and feet blackish green in front, black below (♀ killed in July).

Ceylon, India, China, Formosa, Hainan.

302. THALASSIDROMA ——?

A species of Stormy Petrel is often seen in Colombo harbour and on the west coast in the bad weather during the south-west monsoon, but no specimen of it has yet been obtained; it has appeared to me to be entirely black, with the exception of the white rump.

303. CROICOCEPHALUS ICHTHYAËTUS, Pall.

Layard mentions having seen a pair of these birds at Pt. Pedro after a severe storm. It appears to be only occasionally seen on the Indian coast.

304. XEMA BRUNNICEPHALA, Jerdon.

This is the only true Gull commonly found in Ceylon. It is very abundant in the north, and is seen at times on all parts of the coast.

Ceylon, India, Pekin?

305. SYLOCHELIDON CASPIA, Lath.

306. GELOCHELIDON ANGLICA, Montagu.

307. HYDROCHELIDON LEUCOPAREIA, Natt.

Hydrochelidon indica, Stephens.

These species I have found common in Ceylon, and I have no doubt of their being resident there. *S. caspia* may be seen at all times of the year, almost invariably in pairs, flying along the shore just outside the line of beach. I have shot *G. anglica* in April, July, and December, but have not met with one in the full breeding-plumage.

308. SEENA AURANTIA, Gray.

Said by Layard to be common. I have not actually identified

this species, but believe I have often seen it near the Aripo pearl-banks.

309. STERNA MELANOGASTER, Temm.

I have frequently seen this Tern near Aripo, and occasionally at Colombo. Layard found it common on some of the inland lakes as well as on the coast.

310. STERNA NIGRA, Linn.

Sterna leucoptera, Temm.

I shot one of a pair of these birds in May 1866. They were fly-ing about over a small tank, not very far from the shore, about six miles from Aripo, and were in rather imperfect plumage, the head and neck being speckled. The characters of the species, however, were unmistakable. My specimen is now in the Colombo Museum. This is the only occasion of this Tern having been recognized in Ceylon; and it has only been recently added by Mr. Hume to the Indian avifauna.

Ceylon, India, China, North Africa, South Europe.

311. STERNA GRACILIS, Gould?

I include, with some doubt, under this heading a Tern shot in July 1869, on the Colombo beach; others of the same kind were killed at the time; and they were all in rather immature plumage. This specimen has been examined by Mr. Howard Saunders and Mr. Gould, and is believed by those gentlemen to be *S. gracilis*, and in that case a visitor to Ceylon during the Australian winter. It had the bill reddish black, irides black, and feet dull fleshy red. *S. gracilis* is allied to *S. hirundo,* but has the bill slighter, the upper tail-coverts grey as on the back and tail, and the whole under surface white.

312. STERNULA SINENSIS, Gmel.?

There is, I think, some doubt about the species to which Layard refers under the name of *S. minuta,* and which he speaks of as fre-quenting the inland lakes, though "most common on tanks and still waters near the sea-shore." I have never succeeded in obtaining a specimen of true *S. minuta;* and Mr. Legge, who has collected many of the Ceylon Terns, has been equally unsuccessful; but we have both frequently met with a small species in winter dress which may have been mistaken for it. This bird agrees in measurements and general colouring with *S. sinensis,* Gmel. (*S. sumatrana,* Raffles), and differs from *S. minuta* in having a black bill and the shaft of the first primary white. It was also collected by Mr. Jesse during the late Abyssinian expedition.

313. THALASSEUS CRISTATUS, Stephens.

This Tern is not uncommon on the west coast during summer. I have identified a specimen killed on the beach at Colombo in com-pany with smaller species.

314. THALASSEUS MEDIUS, Horsf.

Thalasseus bengalensis, Less., apud Jerdon.

A Tern apparently of this species is very common.

315. ONYCHOPRION ANÆSTHETUS, Scop.

I do not know this Tern; but Layard mentions having obtained three specimens.

316. PHAËTON RUBRICAUDA, Bodd.

During my annual cruises on the Ceylon coast, I have seen this bird sufficiently near to identify it with certainty, as it hovered over the vessel. All the Tropic birds I have seen there, however, have had white tails; and, as I find among my notes mention of one instance of the bill being red, I conclude that bird was an immature example of the above species. I am very confident I have also seen the yellow-billed species, *P. flavirostris,* Brandt, but I have no special record of the colour of the bill. I shall therefore only call the attention of future observers to that species.

317. SULA FIBER, Linn.

In February and March 1868 I had many opportunities of watching a pair of Boobies which frequented the neighbourhood of the Aripo pearl-banks, about ten miles from the land. They used often to perch on a large iron buoy close to my usual anchorage at night. I only saw them during that one season; and they have not been otherwise recorded.

318. ATTAGEN MINOR, Gmelin.

Attagen ariel, Gould.

Frigate-birds have been killed in several localities on the west coast; and I have observed them on many occasions at Aripo during the strength of the south-west monsoon. They were generally in parties of five or six, and at a considerable height above the shore. Their action, as they hung as it were against the gale, slowly swaying, first on one side, then on the other, strongly reminded me of the behaviour of a large paper kite when it has mounted high in the air. Without any perceptible movement of their partially extended wings, these birds remained as if suspended in the air, but very slowly working against the wind, and gradually advancing along the line of beach. Layard mentions these birds under the name of *A. ariel,* Gould, a species from the Australian seas, but which also has been recorded by Swinhoe from Amoy.

A. aquilus, Linn., is found in the Indian seas; and it is not unlikely that some of the Frigate-birds seen on the Ceylon coast may belong to that species.

319. PELECANUS PHILIPPENSIS, Gmel.

I have seen Pelicans near Trincomalie, and at the entrance to Kokeley Lake, on the north-east coast. Their breeding-stations

are in that part of the island; and they do not appear to wander far away.
Ceylon, India eastward.

320. GRACULUS SINENSIS, Shaw.

Recorded by Layard; I have not identified it.
Ceylon, India, Eastern Asia.

321. GRACULUS JAVANICUS, Horsf.

Very numerous in backwaters along the coast and in lakes inland. It may be seen in dozens perched on the stakes of the fishing-kraals, and will generally allow a boat or canoe to approach within a short distance.
Ceylon, India, Malaya.

322. PLOTUS MELANOGASTER, Forst.

I have seen this bird frequently at Aripo; and it is common on some of the large inland tanks. It is also sometimes found near Colombo.
This species was first described and figured from Ceylon.
Ceylon, India, Burmah, Malaya.

Addendum.

323. PRIONOCHILUS VINCENS, Sclater*.

Discovered by Mr. Vincent Legge, R.A., at the foot of the hills in the south of the island. It is described as frequenting the creeping plants entwining the trunks of the trees. The discovery of this new species in Ceylon is of considerable interest, as it is quite a Malay form, and no representative of the genus has yet been found in India. *Dicæum* is its nearest ally in Ceylon.
"Bill black, paler below; irides reddish; feet brownish black."
Ceylon.

7. Notes on a New Species of Tapir (*Tapirus leucogenys*) from the Snowy Regions of the Cordilleras of Ecuador, and on the Young Spotted Tapirs of Tropical America. By Dr. J. E. GRAY, F.R.S. &c.

[Received February 21, 1872.]

(Plates XXI. & XXII.)

The British Museum has lately received the skins and skeletons of seven Tapirs collected by Mr. Buckley in Ecuador, as under:—
1 & 11. An adult female and a nearly adult male with rather long hair, from Sunia, part of the snowy range of the Cordilleras.

* See below, P. Z. S. for June 18.—P. L. S.

www.ingramcontent.com/pod-product-compliance
Lightning Source LLC
Chambersburg PA
CBHW021956190326
41519CB00009B/1280